Data Ingestion with Python Cookbook

A practical guide to ingesting, monitoring, and identifying errors in the data ingestion process

Gláucia Esppenchutz

BIRMINGHAM—MUMBAI

Data Ingestion with Python Cookbook

Group Product Manager: Reshma Raman

Publishing Product Manager: Arindam Majumdar

Senior Editor: Tiksha Lad

Technical Editor: Devanshi Ayare

Copy Editor: Safis Editing

Project Coordinator: Farheen Fathima

Proofreader: Safis Editing

Indexer: Sejal Dsilva

Production Designer: Jyoti Chauhan

Marketing Coordinator: Nivedita Singh

First published: May 2023

Production reference: 1300523

Published by Packt Publishing Ltd.
Livery Place
35 Livery Street
Birmingham
B3 2PB, UK.

ISBN 978-1-83763-260-2

www.packtpub.com

This book represents a lot and wouldn't be possible without my loving husband, Lincoln, and his support and understanding during this challenging endeavor. I want to thank all my friends that didn't let me give up and always boosted my spirits, along with my grandmother, who always believed, helped, and said I would do big things one day. Finally, I want to thank my beloved and four-pawed best friend, who is at peace, Minduim, for "helping" me to write this book.

– Gláucia Esppenchutz

Contributors

About the author

Gláucia Esppenchutz is a data engineer with expertise in managing data pipelines and vast amounts of data using cloud and on-premises technologies. She worked in companies such as Globo.com, BMW Group, and Cloudera. Currently, she works at AiFi, specializing in the field of data operations for autonomous systems.

She comes from the biomedical field and shifted her career ten years ago to chase the dream of working closely with technology and data. She is in constant contact with the open source community, mentoring people and helping to manage projects, and has collaborated with the Apache, PyLadies group, FreeCodeCamp, Udacity, and MentorColor communities.

I want to thank my patient and beloved husband and my friends. Thanks also to my mentors in the Python open source community and the DataBootCamp founders, who guided me at the beginning of my journey.

Thanks to the Packt team, who helped me through some hard times; you were terrific!

About the reviewers

Bitthal Khaitan is currently working as a big data and cloud engineer with CVS Health, a Fortune 4 organization. He has a demonstrated history of working in the cloud, data and analytics industry for 12+ years. His primary certified skills are **Google Cloud Platform** (**GCP**), the big data ecosystem (Hadoop, Spark, etc.), and data warehousing on Teradata. He has worked in all phases of the SDLC of DW/BI and big data projects with strong expertise in the USA healthcare, insurance and retail domains. He actively helps new graduates with mentoring, resume reviews, and job hunting tips in the data engineering domain. Over 20,000 people follow Bitthal on LinkedIn. He is currently based out of Dallas, Texas, USA.

Jagjeet Makhija is a highly accomplished technology leader with over 20 years of experience. They are skilled not only in various domains including AI, data warehouse architecture, and business analytics, but also have a strong passion for staying ahead of technology trends such as AI and ChatGPT. Jagjeet is recognized for their significant contributions to the industry, particularly in complex proof of concepts and integrating Microsoft products with ChatGPT. They are also an avid book reviewer and have actively shared their extensive knowledge and expertise through presentations, blog articles, and online forums.

Krishnan Raghavan is an IT professional with over 20 years of experience in the area of software development and delivery excellence across multiple domains and technology, ranging from C++ to Java, Python, data warehousing, and big data tools and technologies. Krishnan tries to give back to the community by being part of GDG – Pune Volunteer Group, helping the team in organizing events. When not working, Krishnan likes to spend time with his wife and daughter, as well as reading fiction, non-fiction, and technical books. Currently, he is unsuccessfully trying to learn how to play the guitar.

You can connect with Krishnan at mail to: `krishnan@gmail.com` or via LinkedIn: `www.linkedin.com/in/krishnan-raghavan`

I would like to thank my wife, Anita, and daughter, Ananya, for giving me the time and space to review this book.

Table of Contents

2

Principals of Data Access – Accessing Your Data   31

3

Data Discovery – Understanding Our Data before Ingesting It   71

4

Reading CSV and JSON Files and Solving Problems   95

5

Ingesting Data from Structured and Unstructured Databases   119

6

Using PySpark with Defined and Non-Defined Schemas   159

7

Ingesting Analytical Data   181

Part 2: Structuring the Ingestion Pipeline

8

Designing Monitored Data Workflows   213

9

Putting Everything Together with Airflow   243

10

Logging and Monitoring Your Data Ingest in Airflow   281

11

Automating Your Data Ingestion Pipelines   325

12

Using Data Observability for Debugging, Error Handling, and Preventing Downtime   349

Preface

Welcome to *Data Ingestion with Python Cookbook*. I hope you are excited as me to enter the world of data engineering.

Data Ingestion with Python Cookbook is a practical guide that will empower you to design and implement efficient data ingestion pipelines. With real-world examples and renowned open-source tools, this book addresses your queries and hurdles head-on.

Beginning with designing pipelines, you'll explore working with and without data schemas, constructing monitored workflows using Airflow, and embracing data observability principles while adhering to best practices. Tackling the challenges of reading diverse data sources and formats, you'll gain a comprehensive understanding of all these.

Our journey continues with essential insights into error logging, identification, resolution, data orchestration, and effective monitoring. You'll discover optimal approaches for storing logs, ensuring easy access and references for them in the future.

By the end of this book, you'll possess a fully automated setup to initiate data ingestion and pipeline monitoring. This streamlined process will seamlessly integrate into the subsequent stages of the **Extract, Transform, and Load** (ETL) process, propelling your data integration capabilities to new heights. Get ready to embark on an enlightening and transformative data ingestion journey.

Who this book is for

This comprehensive book is specifically designed for Data Engineers, Data Integration Specialists, and passionate data enthusiasts seeking a deeper understanding of data ingestion processes, data flows, and the typical challenges encountered along the way. It provides valuable insights, best practices, and practical knowledge to enhance your skills and proficiency in handling data ingestion tasks effectively.

Whether you are a beginner in the data world or an experienced developer, this book will suit you. It is recommended to know the Python programming fundamentals and have basic knowledge of Docker to read and run this book's code.

What this book covers

Chapter 1, Introduction to Data Ingestion, introduces you to data ingestion best practices and the challenges of working with diverse data sources. It explains the importance of the tools covered in the book, presents them, and provides installation instructions.

Chapter 2, Data Access Principals – Accessing your Data, explores data access concepts related to data governance, covering workflows and management of familiar sources such as SFTP servers, APIs, and cloud providers. It also provides examples of creating data access policies in databases, data warehouses, and the cloud.

Chapter 3, Data Discovery – Understanding Our Data Before Ingesting It, teaches you the significance of carrying out the data discovery process before data ingestion. It covers manual discovery, documentation, and using an open-source tool, OpenMetadata, for local configuration.

Chapter 4, Reading CSV and JSON Files and Solving Problems, introduces you to ingesting CSV and JSON files using Python and PySpark. It demonstrates handling varying data volumes and infrastructures while addressing common challenges and providing solutions.

Chapter 5, Ingesting Data from Structured and Unstructured Databases, covers fundamental concepts of relational and non-relational databases, including everyday use cases. You will learn how to read and handle data from these models, understand vital considerations, and troubleshoot potential errors.

Chapter 6, Using PySpark with Defined and Non-Defined Schemas, delves deeper into common PySpark use cases, focusing on handling defined and non-defined schemas. It also explores reading and understanding complex logs from Spark (PySpark core) and formatting techniques for easier debugging.

Chapter 7, Ingesting Analytical Data, introduces you to analytical data and common formats for reading and writing. It explores reading partitioned data for improved performance and discusses Reverse ETL theory with real-life application workflows and diagrams.

Chapter 8, Designing Monitored Data Workflows, covers logging best practices for data ingestion, facilitating error identification, and debugging. Techniques such as monitoring file size, row count, and object count enable improved monitoring of dashboards, alerts, and insights.

Chapter 9, Putting Everything Together with Airflow, consolidates the previously presented information and guides you in building a real-life data ingestion application using Airflow. It covers essential components, configuration, and issue resolution in the process.

Chapter 10, Logging and Monitoring Your Data Ingest in Airflow, explores advanced logging and monitoring in data ingestion with Airflow. It covers creating custom operators, setting up notifications, and monitoring for data anomalies. Configuration of notifications for tools such as Slack is also covered to stay updated on the data ingestion process.

Chapter 11, Automating Your Data Ingestion Pipelines, focuses on automating data ingests using previously learned best practices, enabling reader autonomy. It addresses common challenges with schedulers or orchestration tools and provides solutions to avoid problems in production clusters.

Chapter 12, Using Data Observability for Debugging, Error Handling, and Preventing Downtime, explores data observability concepts, popular monitoring tools such as Grafana, and best practices for log storage and data lineage. It also covers creating visualization graphs to monitor data source issues using Airflow configuration and data ingestion scripts.

To get the most out of this book

To execute the code in this book, you must have at least a basic knowledge of Python. We will use Python as the core language to execute the code. The code examples have been tested using Python 3.8. However, it is expected to still work with future language versions.

Along with Python, this book uses Docker to emulate data systems and applications in our local machine, such as PostgreSQL, MongoDB, and Airflow. Therefore, a basic knowledge of Docker is recommended to edit container image files and run and stop containers.

Please, remember that some command-line commands may need adjustments depending on your local settings or operating system. The commands in the code examples are based on the Linux command-line syntax and might need some adaptations to run on Windows PowerShell.

Software/Hardware covered in the book	OS Requirements
Python 3.8 or higher	Windows, Mac OS X, and Linux (any)
Docker Engine 24.0 / Docker Desktop 4.19	Windows, Mac OS X, and Linux (any)

For almost all recipes in this book, you can use a Jupyter Notebook to execute the code. Even though it is not mandatory to install it, this tool can help you to test the code and try new things on the code due to the friendly interface.

If you are using the digital version of this book, we advise you to type the code yourself or access the code via the GitHub repository (link available in the next section). Doing so will help you avoid any potential errors related to the copying and pasting of code.

Download the example code files

You can download the example code files for this book from GitHub at https://github.com/ PacktPublishing/Data-Ingestion-with-Python-Cookbook. In case there's an update to the code, it will be updated on the existing GitHub repository.

We also have other code bundles from our rich catalog of books and videos available at https:// github.com/PacktPublishing/. Check them out!

Download the color images

We also provide a PDF file that has color images of the screenshots/diagrams used in this book. You can download it here: https://packt.link/xwl0U

Conventions used

There are a number of text conventions used throughout this book.

`Code in text`: Indicates code words in text, database table names, folder names, filenames, file extensions, pathnames, dummy URLs, user input, and Twitter handles. Here is an example: "Then we proceeded with the `with open` statement."

A block of code is set as follows:

```
def gets_csv_first_line (csv_file):
    logging.info(f"Starting function to read first line")
    try:
        with open(csv_file, 'r') as file:
            logging.info(f"Reading file")
```

Any command-line input or output is written as follows:

```
$ python3 --version
Python 3.8.10
```

Bold: Indicates a new term, an important word, or words that you see onscreen. For example, words in menus or dialog boxes appear in the text like this. Here is an example: "Then, when we selected `showString` at `NativeMethodAccessorImpl.java:0`, which redirected us to the **Stages** page."

> **Tips or important notes**
> Appear like this.

Sections

In this book, you will find several headings that appear frequently (*Getting ready*, *How to do it...*, *How it works...*, *There's more...*, and *See also*).

To give clear instructions on how to complete a recipe, use these sections as follows:

Getting ready

This section tells you what to expect in the recipe and describes how to set up any software or any preliminary settings required for the recipe.

How to do it...

This section contains the steps required to follow the recipe.

How it works...

This section usually consists of a detailed explanation of what happened in the previous section.

There's more...

This section consists of additional information about the recipe in order to make you more knowledgeable about the recipe.

See also

This section provides helpful links to other useful information for the recipe.

Get in touch

Feedback from our readers is always welcome.

General feedback: If you have questions about any aspect of this book, mention the book title in the subject of your message and email us at customercare@packtpub.com.

Errata: Although we have taken every care to ensure the accuracy of our content, mistakes do happen. If you have found a mistake in this book, we would be grateful if you would report this to us. Please visit www.packtpub.com/support/errata, selecting your book, clicking on the Errata Submission Form link, and entering the details.

Piracy: If you come across any illegal copies of our works in any form on the Internet, we would be grateful if you would provide us with the location address or website name. Please contact us at copyright@packt.com with a link to the material.

If you are interested in becoming an author: If there is a topic that you have expertise in and you are interested in either writing or contributing to a book, please visit authors.packtpub.com.

Share Your Thoughts

Once you've read *Data Ingestion with Python Cookbook*, we'd love to hear your thoughts! Scan the QR code below to go straight to the Amazon review page for this book and share your feedback.

https://packt.link/r/183763260X

Your review is important to us and the tech community and will help us make sure we're delivering excellent quality content.

Download a free PDF copy of this book

Thanks for purchasing this book!

Do you like to read on the go but are unable to carry your print books everywhere?

Is your eBook purchase not compatible with the device of your choice?

Don't worry, now with every Packt book you get a DRM-free PDF version of that book at no cost.

Read anywhere, any place, on any device. Search, copy, and paste code from your favorite technical books directly into your application.

The perks don't stop there, you can get exclusive access to discounts, newsletters, and great free content in your inbox daily

Follow these simple steps to get the benefits:

1. Scan the QR code or visit the link below

https://packt.link/free-ebook/9781837632602

2. Submit your proof of purchase

3. That's it! We'll send your free PDF and other benefits to your email directly

Part 1:
Fundamentals
of Data Ingestion

In this part, you will be introduced to the fundamentals of data ingestion and data engineering, passing through the basic definition of an ingestion pipeline, the common types of data sources, and the technologies involved.

This part has the following chapters:

- *Chapter 1, Introduction to Data Ingestion*
- *Chapter 2, Principals of Data Access – Accessing Your Data*
- *Chapter 3, Data Discovery – Understanding Our Data Before Ingesting It*
- *Chapter 4, Reading CSV and JSON Files and Solving Problems*
- *Chapter 5, Ingesting Data from Structured and Unstructured Databases*
- *Chapter 6, Using PySpark with Defined and Non-Defined Schemas*
- *Chapter 7, Ingesting Analytical Data*

1

Introduction to Data Ingestion

Welcome to the fantastic world of data! Are you ready to embark on a thrilling journey into data ingestion? If so, this is the perfect book to start! Ingesting data is the first step into the big data world.

Data ingestion is a process that involves gathering and importing data and also storing it properly so that the subsequent **extract, transform, and load** (ETL) pipeline can utilize the data. To make it happen, we must be cautious about the tools we will use and how to configure them properly.

In our book journey, we will use **Python** and **PySpark** to retrieve data from different data sources and learn how to store them properly. To orchestrate all this, the basic concepts of **Airflow** will be implemented, along with efficient monitoring to guarantee that our pipelines are covered.

This chapter will introduce some basic concepts about data ingestion and how to set up your environment to start the tasks.

In this chapter, you will build and learn the following recipes:

- Setting up Python and the environment
- Installing PySpark
- Configuring Docker for MongoDB
- Configuring Docker for Airflow
- Logging libraries
- Creating schemas
- Applying data governance in ingestion
- Implementing data replication

Technical requirements

The commands inside the recipes of this chapter use Linux syntax. If you don't use a Linux-based system, you may need to adapt the commands:

- Docker or Docker Desktop
- The SQL client of your choice (recommended); we recommend DBeaver, since it has a community-free version

You can find the code from this chapter in this GitHub repository: `https://github.com/PacktPublishing/Data-Ingestion-with-Python-Cookbook`.

> **Note**
>
> Windows users might get an error message such as **Docker Desktop requires a newer WSL kernel version.** This can be fixed by following the steps here: `https://docs.docker.com/desktop/windows/wsl/`.

Setting up Python and its environment

In the data world, languages such as **Java**, **Scala**, or **Python** are commonly used. The first two languages are used due to their compatibility with the big data tools environment, such as **Hadoop** and **Spark**, the central core of which runs on a **Java Virtual Machine** (**JVM**). However, in the past few years, the use of Python for data engineering and data science has increased significantly due to the language's versatility, ease of understanding, and many open source libraries built by the community.

Getting ready

Let's create a folder for our project:

1. First, open your system command line. Since I use the **Windows Subsystem for Linux** (**WSL**), I will open the WSL application.

2. Go to your home directory and create a folder as follows:

   ```
   $ mkdir my-project
   ```

3. Go inside this folder:

   ```
   $ cd my-project
   ```

4. Check your Python version on your operating system as follows:

   ```
   $ python --version
   ```

Depending on your operational system, you might or might not have output here – for example, WSL 20.04 users might have the following output:

```
Command 'python' not found, did you mean:
 command 'python3' from deb python3
 command 'python' from deb python-is-python3
```

If your Python path is configured to use the `python` command, you will see output similar to this:

```
Python 3.9.0
```

Sometimes, your Python path might be configured to be invoked using `python3`. You can try it using the following command:

```
$ python3 --version
```

The output will be similar to the `python` command, as follows:

```
Python 3.9.0
```

5. Now, let's check our `pip` version. This check is essential, since some operating systems have more than one Python version installed:

    ```
    $ pip --version
    ```

    You should see similar output:

```
pip 20.0.2 from /usr/lib/python3/dist-packages/pip (python 3.9)
```

If your **operating system (OS)** uses a Python version below `3.8x` or doesn't have the language installed, proceed to the *How to do it* steps; otherwise, you are ready to start the following *Installing PySpark* recipe.

How to do it...

We are going to use the official installer from Python.org. You can find the link for it here: `https://www.python.org/downloads/`:

> **Note**
>
> For Windows users, it is important to check your OS version, since Python 3.10 may not be yet compatible with Windows 7, or your processor type (32-bits or 64-bits).

1. Download one of the stable versions.

    At the time of writing, the stable recommended versions compatible with the tools and resources presented here are `3.8`, `3.9`, and `3.10`. I will use the `3.9` version and download it using the following link: `https://www.python.org/downloads/release/python-390/`. Scrolling down the page, you will find a list of links to Python installers according to OS, as shown in the following screenshot.

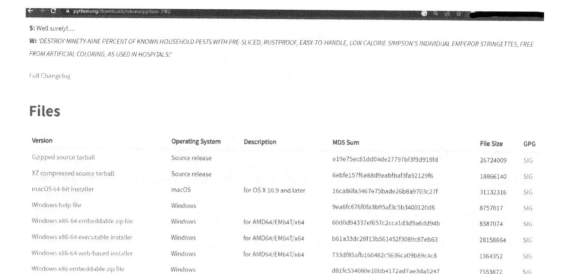

S: Well surely!....

W: 'DESTROY NINETY-NINE PERCENT OF KNOWN HOUSEHOLD PESTS WITH PRE-SLICED, RUSTPROOF, EASY-TO-HANDLE, LOW CALORIE SIMPSON'S INDIVIDUAL EMPEROR STRINGETTES, FREE FROM ARTIFICIAL COLORING, AS USED IN HOSPITALS!'

Full Changelog

Files

Version	Operating System	Description	MD5 Sum	File Size	GPG
Gzipped source tarball	Source release		e19e75ec81dd04de27797bf3f9d918fd	26724009	SIG
XZ compressed source tarball	Source release		6ebfe157f6e88d9eabfbaf3fa92129f6	18866140	SIG
macOS 64-bit installer	macOS	for OS X 10.9 and later	16ca86fa3467e75bade26b8a9703c27f	31132316	SIG
Windows help file	Windows		9ea6fc676f0fa3b95af3c5b3400120d6	8757017	SIG
Windows x86-64 embeddable zip file	Windows	for AMD64/EM64T/x64	60d0d94337ef657c2cca1d3d9a6dd94b	8387074	SIG
Windows x86-64 executable installer	Windows	for AMD64/EM64T/x64	b61a33dc28f13b561452f3089c87eb63	28158664	SIG
Windows x86-64 web-based installer	Windows	for AMD64/EM64T/x64	733df85afb160482c5636ca09b89c4c8	1364352	SIG
Windows x86 embeddable zip file	Windows		d81fc534080e10bb4172ad7ae3da5247	7553872	SIG
Windows x86 executable installer	Windows		4a2812db8ab9f2e522c96c7728cfcccb	27066912	SIG
Windows x86 web-based installer	Windows		cdbfa799e6760c13d06d0c2374110aa3	1327384	SIG

Figure 1.1 – Python.org download files for version 3.9

2. After downloading the installation file, double-click it and follow the instructions in the wizard window. To avoid complexity, choose the recommended settings displayed.

The following screenshot shows how it looks on Windows:

Figure 1.2 – The Python Installer for Windows

3. If you are a Linux user, you can install it from the source using the following commands:

```
$ wget https://www.python.org/ftp/python/3.9.1/Python-3.9.1.tgz

$ tar -xf Python-3.9.1.tgz

$ ./configure –enable-optimizations

$ make -j 9
```

After installing Python, you should be able to execute the `pip` command. If not, refer to the `pip` official documentation page here: `https://pip.pypa.io/en/stable/installation/`.

How it works...

Python is an **interpreted language**, and its interpreter extends several functions made with **C** or **C++**. The language package also comes with several built-in libraries and, of course, the interpreter.

The interpreter works like a Unix shell and can be found in the `usr/local/bin` directory: `https://docs.python.org/3/tutorial/interpreter.html`.

Lastly, note that many Python third-party packages in this book require the `pip` command to be installed. This is because `pip` (an acronym for **Pip Installs Packages**) is the default package manager for Python; therefore, it is used to install, upgrade, and manage the Python packages and dependencies from the **Python Package Index (PyPI)**.

There's more...

Even if you don't have any Python versions on your machine, you can still install them using the command line or **HomeBrew** (for **macOS** users). Windows users can also download them from the MS Windows Store.

Note

If you choose to download Python from the Windows Store, ensure you use an application made by the Python Software Foundation.

See also

You can use `pip` to install convenient third-party applications, such as Jupyter. This is an open source, web-based, interactive (and user-friendly) computing platform, often used by data scientists and data engineers. You can install it from the official website here: `https://jupyter.org/install`.

Installing PySpark

To process, clean, and transform vast amounts of data, we need a tool that provides resilience and distributed processing, and that's why **PySpark** is a good fit. It gets an API over the Spark library that lets you use its applications.

Getting ready

Before starting the PySpark installation, we need to check our Java version in our operational system:

1. Here, we check the Java version:

    ```
    $ java -version
    ```

    You should see output similar to this:

    ```
    openjdk version "1.8.0_292"
    OpenJDK Runtime Environment (build 1.8.0_292-8u292-b10-
    0ubuntu1~20.04-b10)
    OpenJDK 64-Bit Server VM (build 25.292-b10, mixed mode)
    ```

    If everything is correct, you should see the preceding message as the output of the command and the **OpenJDK 18** version or higher. However, some systems don't have any Java version installed by default, and to cover this, we need to proceed to *step 2*.

2. Now, we download the **Java Development Kit (JDK)**.

    Go to https://www.oracle.com/java/technologies/downloads/, select your **OS**, and download the most recent version of JDK. At the time of writing, it is JDK 19.

    The download page of the JDK will look as follows:

JDK 17 will receive updates under these terms, until at least September 2024.

Java 19   Java 17

Java SE Development Kit 19 downloads

Thank you for downloading this release of the Java™ Platform, Standard Edition Development Kit (JDK™). The JDK is a development environment for building applications and components using the Java programming language.

The JDK includes tools for developing and testing programs written in the Java programming language and running on the Java platform.

Linux   macOS   Windows

Product/file description	File size	Download
Arm 64 Compressed Archive	179.64 MB	https://download.oracle.com/java/19/latest/jdk-19_linux-aarch64_bin.tar.gz (sha256)
Arm 64 RPM Package	159.76 MB	https://download.oracle.com/java/19/latest/jdk-19_linux-aarch64_bin.rpm (sha256)
x64 Compressed Archive	180.86 MB	https://download.oracle.com/java/19/latest/jdk-19_linux-x64_bin.tar.gz (sha256)
x64 Debian Package	154.59 MB	https://download.oracle.com/java/19/latest/jdk-19_linux-x64_bin.deb (sha256)
x64 RPM Package	161.40 MB	https://download.oracle.com/java/19/latest/jdk-19_linux-x64_bin.rpm (sha256)

Figure 1.3 – The JDK 19 downloads official web page

Execute the downloaded application. Click on the application to start the installation process. The following window will appear:

> **Note**
> Depending on your OS, the installation window may appear slightly different.

Welcome to the Installation Wizard for Java SE Development Kit 19

This wizard will guide you through the installation process for the Java SE Development Kit 19.

Next >   Cancel

Figure 1.4 – The Java installation wizard window

Click **Next** for the following two questions, and the application will start the installation. You don't need to worry about where the JDK will be installed. By default, the application is configured, as standard, to be compatible with other tools' installations.

3. Next, we again check our Java version. When executing the command again, you should see the following version:

```
$ java -version
openjdk version "1.8.0_292"
OpenJDK Runtime Environment (build 1.8.0_292-8u292-b10-
0ubuntu1~20.04-b10)
OpenJDK 64-Bit Server VM (build 25.292-b10, mixed mode)
```

How to do it...

Here are the steps to perform this recipe:

1. Install PySpark from PyPi:

```
$ pip install pyspark
```

If the command runs successfully, the installation output's last line will look like this:

```
Successfully built pyspark
Installing collected packages: py4j, pyspark
Successfully installed py4j-0.10.9.5 pyspark-3.3.2
```

2. Execute the `pyspark` command to open the interactive shell. When executing the `pyspark` command in your command line, you should see this message:

```
$ pyspark
Python 3.8.10 (default, Jun 22 2022, 20:18:18)
[GCC 9.4.0] on linux
Type "help", "copyright", "credits" or "license" for more
information.
22/10/08 15:06:11 WARN Utils: Your hostname, DESKTOP-DVUDB98
resolves to a loopback address: 127.0.1.1; using 172.29.214.162
instead (on interface eth0)
22/10/08 15:06:11 WARN Utils: Set SPARK_LOCAL_IP if you need to
bind to another address
22/10/08 15:06:13 WARN NativeCodeLoader: Unable to load native-
hadoop library for your platform... using builtin-java classes
where applicable
Using Spark's default log4j profile: org/apache/spark/log4j-
defaults.properties
Setting default log level to "WARN".
To adjust logging level use sc.setLogLevel(newLevel). For
SparkR, use setLogLevel(newLevel).
Welcome to

      ____              __
     / __/__  ___ _____/ /__
    _\ \/ _ \/ _ `/ __/  '_/
   /__ / .__/\_,_/_/ /_/\_\   version 3.1.2
      /_/

Using Python version 3.8.10 (default, Jun 22 2022 20:18:18)
Spark context Web UI available at http://172.29.214.162:4040
Spark context available as 'sc' (master = local[*], app id =
local-1665237974112).
SparkSession available as 'spark'.
>>>
```

You can observe some interesting messages here, such as the Spark version and the Python used from PySpark.

3. Finally, we exit the interactive shell as follows:

```
>>> exit()
$
```

How it works...

As seen at the beginning of this recipe, Spark is a robust framework that runs on top of the JVM. It is also an open source tool for creating resilient and distributed processing output from vast data. With the growth in popularity of the Python language in the past few years, it became necessary to have a solution that adapts Spark to run alongside Python.

PySpark is an interface that interacts with **Spark APIs via Py4J**, dynamically allowing Python code to interact with the JVM. We first need to have Java installed on our OS to use Spark. When we install PySpark, it already comes with Spark and Py4J components installed, making it easy to start the application and build the code.

There's more...

Anaconda is a convenient way to install PySpark and other data science tools. This tool encapsulates all manual processes and has a friendly interface for interacting with and installing Python components, such as **NumPy**, **pandas**, or **Jupyter**:

1. To install Anaconda, go to the official website and select **Products | Anaconda Distribution**: `https://www.anaconda.com/products/distribution`.

2. Download the distribution according to your OS.

For more detailed information about how to install Anaconda and other powerful commands, refer to `https://docs.anaconda.com/`.

Using virtualenv with PySpark

It is possible to configure and use `virtualenv` with PySpark, and Anaconda does it automatically if you choose this type of installation. However, for the other installation methods, we need to make some additional steps to make our Spark cluster (locally or on the server) run it, which includes indicating the `virtualenv /bin/` folder and where your PySpark path is.

See also

There is a nice article about this topic, *Using VirtualEnv with PySpark*, by jzhang, here: `https://community.cloudera.com/t5/Community-Articles/Using-VirtualEnv-with-PySpark/ta-p/245932`.

Configuring Docker for MongoDB

MongoDB is a **Not Only SQL** (**NoSQL**) document-oriented database, widely used to store **Internet of Things** (**IoT**) data, application logs, and so on. A NoSQL database is a non-relational database that stores unstructured data differently from relational databases such as MySQL or PostgreSQL. Don't worry too much about this now; we will cover it in more detail in *Chapter 5*.

Your cluster production environment can handle huge amounts of data and create resilient data storage.

Getting ready

Following the good practice of code organization, let's start creating a folder inside our project to store the Docker image:

Create a folder inside our project directory to store the MongoDB Docker image and data as follows:

```
my-project$ mkdir mongo-local
my-project$ cd mongo-local
```

How to do it...

Here are the steps to try out this recipe:

1. First, we pull the Docker image from Docker Hub as follows:

    ```
    my-project/mongo-local$ docker pull mongo
    ```

    You should see the following message in your command line:

    ```
    Using default tag: latest
    latest: Pulling from library/mongo
    (...)
    bc8341d9c8d5: Pull complete
    (...)
    Status: Downloaded newer image for mongo:latest
    docker.io/library/mongo:latest
    ```

> **Note**
>
> If you are a WSL user, an error might occur if you use the WSL 1 version instead of version 2. You can easily fix this by following the steps here: https://learn.microsoft.com/en-us/windows/wsl/install.

2. Then, we run the MongoDB server as follows:

```
my-project/mongo-local$ docker run \
--name mongodb-local \
-p 27017:27017 \
-e MONGO_INITDB_ROOT_USERNAME="your_username" \
-e MONGO_INITDB_ROOT_PASSWORD="your_password"\
-d mongo:latest
```

We then check our server. To do this, we can use the command line to see which Docker images are running:

```
my-project/mongo-local$ docker ps
```

We then see this on the screen:

Figure 1.5 – MongoDB and Docker running container

We can even check on the Docker Desktop application to see whether our container is running:

Figure 1.6 – The Docker Desktop vision of the MongoDB container running

3. Finally, we need to stop our container. We need to use Container ID to stop the container, which we previously saw when checking the Docker running images. We will rerun it in *Chapter 5*:

```
my-project/mongo-local$ docker stop 427cc2e5d40e
```

How it works...

MongoDB's architecture uses the concept of **distributed processing**, where the main node interacts with clients' requests, such as queries and document manipulation. It distributes the requests automatically among its shards, which are a subset of a larger data collection here.

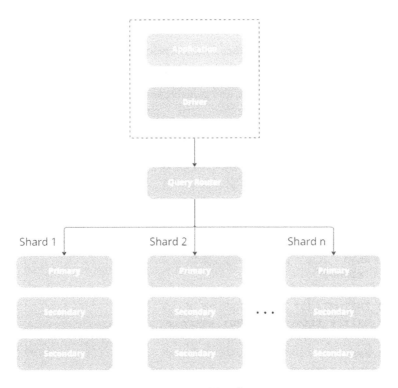

Figure 1.7 – MongoDB architecture

Since we may also have other running projects or software applications inside our machine, isolating any database or application server used in development is a good practice. In this way, we ensure nothing interferes with our local servers, and the debug process can be more manageable.

This Docker image setting creates a MongoDB server locally and even allows us to make additional changes if we want to simulate any other scenario for testing or development.

The commands we used are as follows:

- The --name command defines the name we give to our container.
- The -p command specifies the port our container will open so that we can access it via localhost:27017.
- -e command defines the environment variables. In this case, we set the root username and password for our MongoDB container.
- -d is detached mode – that is, the Docker process will run in the background, and we will not see input or output. However, we can still use docker ps to check the container status.
- mongo:latest indicates Docker pulling this image's latest version.

There's more...

For frequent users, manually configuring other parameters for the MongoDB container, such as the version, image port, database name, and database credentials, is also possible.

A version of this image with example values is also available as a `docker-compose` file in the official documentation here: `https://hub.docker.com/_/mongo`.

The `docker-compose` file for MongoDB looks similar to this:

```
# Use your own values for username and password
version: '3.1'

services:

  mongo:
    image: mongo
    restart: always
    environment:
      MONGO_INITDB_ROOT_USERNAME: root
      MONGO_INITDB_ROOT_PASSWORD: example

  mongo-express:
    image: mongo-express
    restart: always
    ports:
      - 8081:8081
    environment:
      ME_CONFIG_MONGODB_ADMINUSERNAME: root
      ME_CONFIG_MONGODB_ADMINPASSWORD: example
      ME_CONFIG_MONGODB_URL: mongodb://root:example@mongo:27017/
```

See also

You can check out MongoDB at the complete Docker Hub documentation here: `https://hub.docker.com/_/mongo`.

Configuring Docker for Airflow

In this book, we will use **Airflow** to orchestrate data ingests and provide logs to monitor our pipelines.

Airflow can be installed directly on your local machine and any server using PyPi (`https://pypi.org/project/apache-airflow/`) or a Docker container (`https://hub.docker.com/r/apache/airflow`). An official and supported version of Airflow can be found on Docker Hub, and the **Apache Foundation** community maintains it.

However, there are some additional steps to configure our Airflow. Thankfully, the Apache Foundation also has a `docker-compose` file that contains all other requirements to make Airflow work. We just need to complete a few more steps.

Getting ready

Let's start by initializing our Docker application on our machine. You can use the desktop version or the CLI command.

Make sure you are inside your project folder for this. Create a folder to store Airflow internal components and the `docker-compose.yaml` file:

```
my-project$ mkdir airflow-local
my-project$ cd airflow-local
```

How to do it...

1. First, we fetch the `docker-compose.yaml` file directly from the Airflow official docs:

    ```
    my-project/airflow-local$ curl -LfO 'https://airflow.apache.org/
    docs/apache-airflow/2.3.0/docker-compose.yaml'
    ```

    You should see output like this:

% Total	% Received % Xferd	Average Speed Dload Upload	Time Total	Time Spent	Time Left	Current Speed
100 9879	100 9879 0	0 98790	0 --:--:--	--:--:--	--:--:--	98790

Figure 1.8 – Airflow container image download progress

> **Note**
> Check the most stable version of this `docker-compose` file when you download it, since new, more appropriate versions may be available after this book is published.

2. Next, we create the `dags`, `logs`, and `plugins` folders as follows:

    ```
    my-project/airflow-local$ mkdir ./dags ./logs ./plugins
    ```

3. Then, we create and set the Airflow user as follows:

    ```
    my-project/airflow-local$ echo -e "AIRFLOW_UID=$(id -u)\
    nAIRFLOW_GID=0" > .env
    ```

> **Note**
>
> If you have any error messages related to the `AIRFLOW_UID` variable, you can create a `.env` file in the same folder where your `docker-compose.yaml` file is and define the variable as `AIRFLOW_UID=50000`.

4. Then, we initialize the database:

```
my-project/airflow-local$ docker-compose up airflow-init
```

After executing the command, you should see output similar to this:

```
Creating network "airflow-local_default" with the default driver
Creating volume "airflow-local_postgres-db-volume" with default
driver
Pulling postgres (postgres:13)...
13: Pulling from library/postgres
(...)
Status: Downloaded newer image for postgres:13
Pulling redis (redis:latest)...
latest: Pulling from library/redis
bd159e379b3b: Already exists
(...)
Status: Downloaded newer image for redis:latest
Pulling airflow-init (apache/airflow:2.3.0)...
2.3.0: Pulling from apache/airflow
42c077c10790: Pull complete
(...)
Status: Downloaded newer image for apache/airflow:2.3.0
Creating airflow-local_postgres_1 ... done
Creating airflow-local_redis_1     ... done
Creating airflow-local_airflow-init_1 ... done
Attaching to airflow-local_airflow-init_1
(...)
airflow-init_1      | [2022-10-09 09:49:26,250] {manager.
py:213} INFO - Added user airflow
airflow-init_1      | User "airflow" created with role "Admin"
(...)
airflow-local_airflow-init_1 exited with code 0
```

5. Then, we start the Airflow service:

```
my-project/airflow-local$ docker-compose up
```

6. Then, we need to check the Docker processes. Using the following CLI command, you will see the Docker images running:

```
my-project/airflow-local$ docker ps
```

These are the images we see:

CONTAINER ID	IMAGE	COMMAND	CREATED	STATUS	PORTS	NAMES
f8af39abbc66	apache/airflow:2.3.0	"/usr/bin/dumb-init …"	10 minutes ago	Up 3 minutes (healthy)	8080/tcp	airflow-local_airflow-worker_1
1c60f7606f7e	apache/airflow:2.3.0	"/usr/bin/dumb-init …"	10 minutes ago	Up 3 minutes (healthy)	0.0.0.0:8080->8080/tcp	airflow-local_airflow-webserver_1
719dd9b1258f	apache/airflow:2.3.0	"/usr/bin/dumb-init …"	10 minutes ago	Up 3 minutes (healthy)	8080/tcp	airflow-local_airflow-triggerer_1
28d1f719cfe8	apache/airflow:2.3.0	"/usr/bin/dumb-init …"	10 minutes ago	Up 3 minutes (healthy)	0.0.0.0:5555->5555/tcp, 8080/tcp	airflow-local_flower_1
2a37aaf567c4	apache/airflow:2.3.0	"/usr/bin/dumb-init …"	10 minutes ago	Up 3 minutes (healthy)	8080/tcp	airflow-local_airflow-scheduler_1
04d5ebba2bc2	postgres:13	"docker-entrypoint.s…"	10 minutes ago	Up 4 minutes (healthy)	5432/tcp	airflow-local_postgres_1
60ac31a7c3ee	redis:latest	"docker-entrypoint.s…"	10 minutes ago	Up 4 minutes (healthy)	6379/tcp	airflow-local_redis_1

Figure 1.9 – The docker ps command output

In the Docker Desktop application, you can also see the same containers running but with a more friendly interface:

Figure 1.10 – A Docker desktop view of the Airflow containers running

7. Then, we access Airflow in a web browser:

In your preferred browser, type `http://localhost:8080/home`. The following screen will appear:

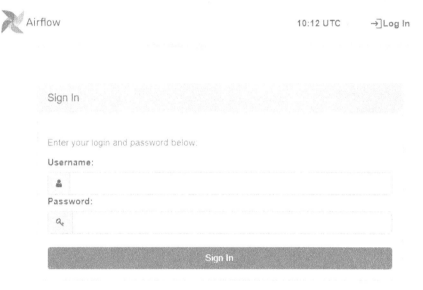

Figure 1.11 – The Airflow UI login page

8. Then, we log in to the Airflow platform. Since it's a local application used for testing and learning, the default credentials (username and password) for administrative access in Airflow are `airflow`.

When logged in, the following screen will appear:

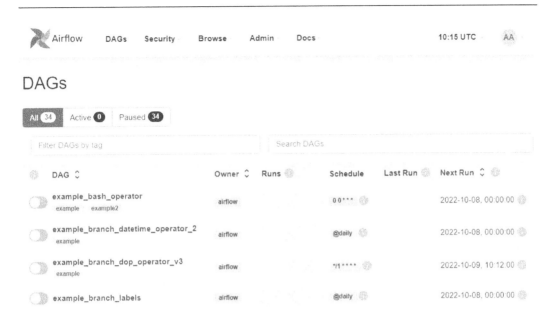

Figure 1.12 – The Airflow UI main page

9. Then, we stop our containers. We can stop our containers until we reach *Chapter 9*, when we will explore data ingest in Airflow:

```
my-project/airflow-local$ docker-compose stop
```

How it works...

Airflow is an open source platform that allows batch data pipeline development, monitoring, and scheduling. However, it requires other components, such as an internal database, to store metadata to work correctly. In this example, we use PostgreSQL to store the metadata and **Redis** to cache information.

All this can be installed directly in our machine environment one by one. Even though it seems quite simple, it may not be due to compatibility issues with OS, other software versions, and so on.

Docker can create an isolated environment and provide all the requirements to make it work. With `docker-compose`, it becomes even simpler, since we can create dependencies between the components that can only be created if the others are healthy.

You can also open the `docker-compose.yaml` file we downloaded for this recipe and take a look to explore it better. We will also cover it in detail in *Chapter 9*.

See also

If you want to learn more about how this `docker-compose` file works, you can look at the Apache Airflow official Docker documentation on the Apache Airflow documentation page: `https://airflow.apache.org/docs/apache-airflow/stable/howto/docker-compose/index.html`.

Creating schemas

Schemas are considered blueprints of a database or table. While some databases strictly require schema definition, others can work without it. However, in some cases, it is advantageous to work with data schemas to ensure that the application data architecture is maintained and can receive the desired data input.

Getting ready

Let's imagine we need to create a database for a school to store information about the students, the courses, and the instructors. With this information, we know we have at least three tables so far.

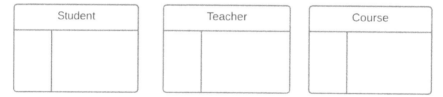

Figure 1.13 – A table diagram for three entities

In this recipe, we will cover how schemas work using the **Entity Relationship Diagram** (**ERD**), a visual representation of relationships between entities in a database, to exemplify how schemas are connected.

How to do it...

Here are the steps to try this:

1.   We define the type of schema. The following figure helps us understand how to go about this:

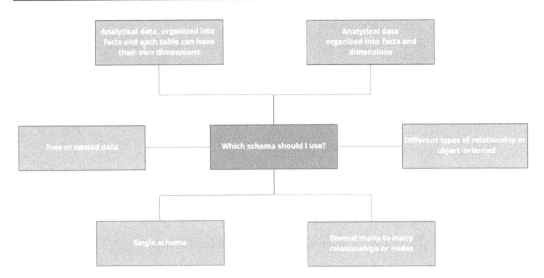

Figure 1.14 – A diagram to help you decide which schema to use

2. Then, we define the fields and the data type for each table column:

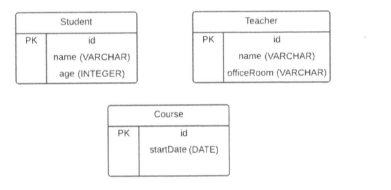

Figure 1.15 – A definition of the columns of each table

3. Next, we define which fields can be empty or NULL:

Student		
PK	id	INTEGER NOT NULL
	name	VARCHAR NOT NULL
	age	INTEGER NOT NULL

Teacher		
PK	id	INTEGER NOT NULL
	name	VARCHAR NOT NULL
	officeRoom	VARCHAR NOT NULL

Course		
PK	id	INTEGER NOT NULL
	startDate	DATE NOT NULL

Figure 1.16 – A definition of which columns can be NULL

4. Then, we create the relationship between the tables:

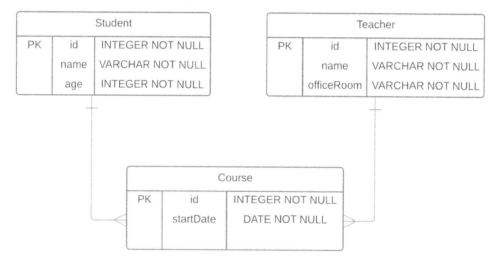

Student		
PK	id	INTEGER NOT NULL
	name	VARCHAR NOT NULL
	age	INTEGER NOT NULL

Teacher		
PK	id	INTEGER NOT NULL
	name	VARCHAR NOT NULL
	officeRoom	VARCHAR NOT NULL

Course		
PK	id	INTEGER NOT NULL
	startDate	DATE NOT NULL

Figure 1.17 – A relationship diagram of the tables

How it works...

When designing data schemas, the first thing we need to do is define their type. As we can see in the diagram in *step 1*, applying the schema architecture depends on the data's purpose.

After that, the tables are designed. Deciding how to define data types can vary, depending project or purpose, but deciding what values a column can receive is important. For instance, the `officeRoom` on `Teacher` table can be an `Integer` type if we know the room's identification is always numeric, or a `String` type if it is unsure how identifications are made (for example, `Room 3-D`).

Another important topic covered in *step 3* is how to define which of the columns can accept NULL fields. Can a field for a student's name be empty? If not, we need to create a constraint to forbid this type of insert.

Finally, based on the type of schema, a definition of the relationship between the tables is made.

See also

If you want to know more about database schema designs and their application, read this article by Mark Smallcombe: `https://www.integrate.io/blog/database-schema-examples/`.

Applying data governance in ingestion

Data governance is a set of methodologies that ensure that data is secure, available, well-stored, documented, private, and accurate.

Getting ready

Data ingestion is the beginning of the data pipeline process, but it doesn't mean data governance is not heavily applied. The governance status in the final data pipeline output depends on how it was implemented during the ingestion.

The following diagram shows how data ingestion is commonly conducted:

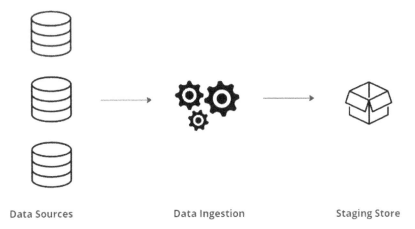

Data Sources   Data Ingestion   Staging Store

Figure 1.18 – The data ingestion process

Let's analyze the steps in the diagram:

1. **Getting data from the source**: The first step is to define the type of data, its periodicity, where we will gather it, and why we need it.

2. **Writing the scripts to ingest data**: Based on the answers to the previous step, we can begin planning how our code will behave and some basic steps.

3. **Storing data in a temporary database or other types of storage**: Between the ingest and the transformation phase, data is typically stored in a temporary database or repository.

Figure 1.19 – Data governance pillars

How to do it...

Step by step, let's attribute the pillars in *Figure 1.19* to the ingestion phase:

1. A concern for accessibility needs to be applied at the data source level, defining the individuals that are allowed to see or retrieve data.

2. Next, it is necessary to catalog our data to understand it better. Since data ingestion is only covered here, it is more relevant to cover the data sources.

3. The quality pillar will be applied to the ingestion and staging area, where we control the data and keep its quality aligned with the source.

4.   Then, let's define ownership. We know the data source *belongs* to a business area or a company. However, when we ingested the data and put it in temporary or staging storage, it becomes our responsibility to maintain it.

5.   The last pillar involves keeping data secure for the whole pipeline. Security is vital in all steps, since we may be handling private or sensitive information.

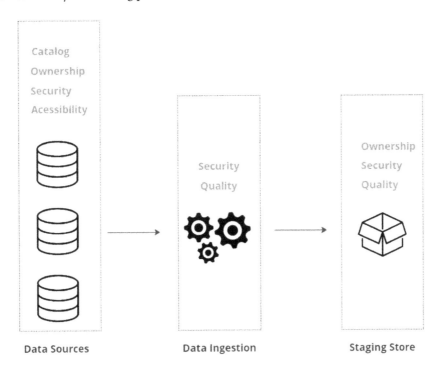

Figure 1.20 – Adding to data ingestion

How it works...

While some articles define "pillars" to create governance good practices, the best way to understand how to apply them is to understand how they are composed. As you saw in the previous *How to do it...* section, we attributed some items to our pipeline, and now we can understand how they are connected to the following topics:

- **Data accessibility**: Data accessibility is how people from a group, organization, or project can see and use data. The information needs to be readily available for use. At the same time, it needs to be available for the people involved in the process. For example, sensitive data accessibility should be restricted to some people or programs. In the diagram we built, we applied it to our data sources, since we need to understand and retrieve data. For the same reason, it can be applied for temporary storage needs as well.

- **Data catalog**: Cataloging and documenting data are essential for business and engineering teams. When we know what types of information rely on our databases or data lakes and have quick access to these documents, the action time to solve a problem becomes short.

  Again, documenting our data sources can make the ingest process quicker, since we need to make a discovery every time we need to ingest data.

- **Data quality**: Quality is constantly preoccupied with ingesting, processing, and loading data. Tracking and monitoring data's expected income and outcome by its periodicity is essential. For example, if we expect to ingest 300 GB of data per day and suddenly it drops to 1 GB, something is very wrong and will affect the quality of our final output. Other quality parameters can be the number of columns, partitioning, and so on, which we will explore later in this book.

- **Ownership**: Who is responsible for the data? This definition is crucial to make contact with the owner if there are problems or attribute responsibility to keep and maintain data.

- **Security**: A concerning topic nowadays is data security. With so many regulations about data privacy, it became an obligation of data engineers and scientists to know, at least, the basics of encryption, sensitive data, and how to avoid data leaks. Even languages and libraries that are used for work need to be evaluated. That's why this item is attributed to the three steps in *Figure 1.19*.

In addition to the topics we explored, a global data governance project has a vital role called a **data steward**, which is responsible for managing an organization's data assets and ensuring that data is accurate, consistent, and secure. In summary, data stewardship is managing and overseeing an organization's data assets.

See also

You can read more about a recent vulnerability found in one of the most used tools for data engineering here: https://www.ncsc.gov.uk/information/log4j-vulnerability-what-everyone-needs-to-know.

Implementing data replication

Data replication is a process applied in data environments to create multiple copies of data and store them on different locations, servers, or sites. This technique is commonly implemented to create better availability and avoid data loss if there is downtime, or even a natural disaster that affects a data center.

Getting ready

You will find across papers and articles different types (or even names) on the best way for **data replication** decision. In this recipe, you will learn how to decide which kind of replication better suits your application or software.

How to do it...

Let's begin to build our fundamental pillars to implement data replication:

1. First, we need to decide the size of our replication, and it can be done using a portion or all the stored data.

2. The next step is to consider when replication will take place. It can be done synchronously when new data arrives in storage or within a specific timeframe.

3. The last fundamental pillar is whether the data is incremented or in a bulk form.

    In the end, we will have a diagram that looks like the following:

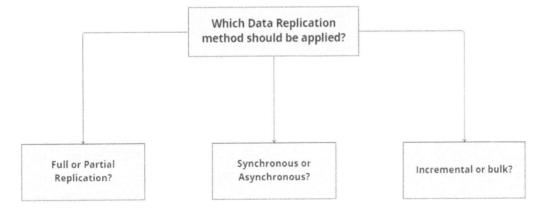

Figure 1.21 – A data replication model decision diagram

How it works...

Analyzing the preceding figure, we have three main questions to answer, regarding the extension, the frequency, and whether our replication will be incremental or bulk.

For the first question, we decide whether the replication will be complete or partial. In other words, either the data will consistently be replicated no matter what type of transaction or change was made, or just a portion of the data will be replicated. A real example of this would be keeping track of all store sales or just the most expensive ones.

The second question, related to the frequency, is to decide when a replication needs to be done. This question also needs to take into consideration related costs. Real-time replication is often more expensive, but the synchronicity guarantees almost no data inconsistency.

Lastly, it is relevant to consider how data will be transported to the replication site. In most cases, a scheduler with a script can replicate small data batches and reduce transportation costs. However, a bulk replication can be used in the data ingestion process, such as copying all the current batch's raw data from a source to cold storage.

There's more...

One method of data replication that has seen an increase in use in the past few years is **cold storage**, which is used to retain data used infrequently or is even inactive. The costs related to this type of replication are meager and guarantee data longevity. You can find cold storage solutions in all cloud providers, such as **Amazon Glacier**, **Azure Cool Blob**, and **Google Cloud Storage Nearline**.

Besides replication, regulatory compliance such as **General Data Protection Regulation** (**GDPR**) laws benefit from this type of storage, since, for some case scenarios, users' data need to be kept for some years.

In this chapter, we explored the basic concepts and laid the foundation for the following chapters and recipes in this book. We started with a Python installation, prepared our Docker containers, and saw data governance and replication concepts. You will observe over the upcoming chapters that almost all topics interconnect, and you will understand the relevance of understanding them at the beginning of the ETL process.

Further reading

- https://www.manageengine.com/device-control/data-replication.html

Principals of Data Access – Accessing Your Data

Data access is a term that refers to the ability to store, retrieve, transfer, and copy data from one system or application to another. It crucially involves security, legal, and, in some cases, national matters. In addition to the last two, we will also cover some security topics in this chapter.

As data engineers or scientists, knowing how to retrieve data correctly is necessary. Some of it may require **encrypted authentication**, and for this, we need to understand how some decrypting libraries work and how to use them without compromising or leaking sensitive data. Data access also refers to the levels of authorization a system or database have, from administration to read-only roles.

In this chapter, we will cover how the levels of data access are defined and the most used libraries and authentication methods in the data ingestion process.

In this chapter, you will work through the following recipes:

- Implementing governance in a data access workflow
- Accessing databases and data warehouses
- Accessing **SSH File Transfer Protocol** (SFTP) files
- Retrieving data using API authentication
- Managing encrypted files
- Accessing data from AWS
- Accessing data from GCP

Technical requirements

A Google Cloud account can be easily created if you already have a Gmail account, and most of the resources can be accessed with a free tier. It also provides $300 of credit for resources that are not free. It is a good incentive if you want to make other tests using the other recipes in this book inside GCP.

To access and enable a Google Cloud account, go to the `https://cloud.google.com/` page and follow the steps provided on the screen.

> **Note**
> All the recipes covered in this chapter are eligible to use the free tier.

You can also find the code from this chapter in this GitHub repository here: `https://github.com/PacktPublishing/Data-Ingestion-with-Python-Cookbook`.

Implementing governance in a data access workflow

As we saw previously, **data access** or **accessibility** is a **governance** pillar and is closely related to security. Data safety is not only a concern for administrators or managers but also for everyone that is involved with data. Having said that, it is essential to know how to design a base workflow to implement security layers for our data, allowing only authorized people to read or manipulate it.

This recipe will create a workflow with essential topics to implement data access management.

Getting ready

Before designing our workflow, we need to identify the vectors interfering with our data access.

So, what are data vectors?

Vectors are paths someone can use to gain unauthorized access to a server, network, or database. In this case, we will identify the ones related to data leaks.

Let's explore them in a visual form, as shown in the following diagram:

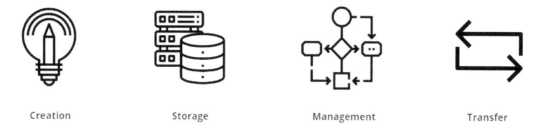

Creation   Storage   Management   Transfer

Figure 2.1 – Data governance vectors

Let us understand each of these stages in the path here:

1. **Data creation**: In this step, we identify where data is created and by who. With this definition, we can ensure only the **accountable** can have access to create or update data.

2. **Storage of data**: After creation, it's important to know where our data is or will be stored. Depending on the answer to this question, the methods to retrieve data will be different and can require additional steps.

3. **Management of users and services**: Data must be used, and people need access. Here, we define the *actors* or the roles we might have, and the common types are **administrator**, **write**, and **read-only** roles.

4. **Transferring data**: How will our data be transferred? It is essential to decide whether it will be real-time, near real-time, or batch. You can add further questions to your workflows, such as how the data will be available for transfers via API or any other method.

How to do it...

After identifying our vectors, we can define the implementation workflow for data access management.

To make it easy to understand how to implement it, let's imagine a hypothetical scenario of a new application where we want to retrieve medical records from patients.

Here is how we do it:

1. The first step is to document all our data and classify it. If there is confidential data, we need to work out how to identify it.

2. Then, we will start to define who can access the data accordingly with the necessary usage. For example, we determine the data administrators and write or read-only permissions here.

3. Once levels of access to data are implemented, we start observing how users will behave. Implementing logs to a database, data warehouse, or any other system with user activity is crucial.

4. Finally, we examine the whole process to determine whether any change is needed.

   In the end, we will have a flow diagram similar to the following diagram:

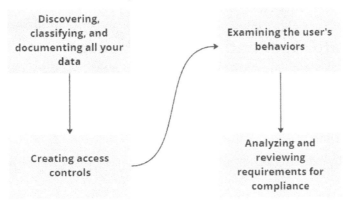

Figure 2.2 – A flow diagram to start implementing data governance

How it works...

Data access management is an ongoing process. Every day, we ingest and create new pipelines to be used by several people on different teams. Here is how it goes:

1. **Discovering, classifying, and documenting all your data**: The first thing to organize is our data. Which patient's data are we going to retrieve? Does it contain **Personally Identifiable Information (PII)** or **Protected/Personal Health Information (PHI)**? Since it's the first time we are ingesting this data, we need to catalog it with flags about PII and who is responsible for the source data.

2. **Creating access controls**: Here, we define access based on roles since not everybody needs access to patient history. We allocate permissions to data based on the roles, responsibilities, and classification.

3. **Examining the users' behaviors**: In this step, we observe how our users behave in their roles. The creation, updates, and deletion actions are logged to be monitored and reviewed if needed. If a medical department no longer uses a report, we can restrict their access or even stop them from ingesting information for them.

4. **Analyzing and reviewing requirements for compliance**: We must ensure our access management follows compliance and local regulations. Legal regulations apply differently for different types of data, which needs to be considered.

See also

- *Healthcare Data Breaches Statistics*: `https://www.hipaajournal.com/healthcare-data-breach-statistics/`

- *European Data Protection Supervisor. Health data in the workspace*: `https://edps.europa.eu/data-protection/data-protection/reference-library/health-data-workplace_en`

Accessing databases and data warehouses

Databases are the foundation of any system or application, no matter your architecture. A database is sometimes needed to store logs, user activities or information, and system stuff.

Putting it in a bigger perspective, data warehouses have the same usage but are related to analytical data. After ingesting and transforming data, we need to load it somewhere where it is easier to retrieve analytic information for use on dashboards, reports, etc.

Currently, it is possible to find several types of databases (of the SQL and NoSQL types) and data warehouse architectures. However, this recipe aims to cover how access control is usually done for both relational structures. The goal is to understand how the access levels are defined, even using a generic scenario.

Getting ready

For this recipe, we will use MySQL. You can install it following the instructions on the MySQL official page here: `https://dev.mysql.com/downloads/installer/`.

You can use any SQL client you choose to execute the queries here. In my case, I will use MySQL Workbench:

1. First, we will create a database using our `root` user. You can name it what you like. My recommendation here is to set the charset to UTF-8:

   ```
   CREATE SCHEMA `your_schema_name` DEFAULT CHARACTER SET utf8 ;
   ```

   My schema will be called `cookbook-data`.

2. The next step is creating tables. Still using the root account, we will create a `people_city` table using the **Data Definition Language** (**DDL**) syntax:

   ```
   CREATE TABLE `cookbook-data`.`people_city` (
     `id` INT NOT NULL,
     `name` VARCHAR(45) NULL,
     `country` VARCHAR(45) NULL,
     `occupation` VARCHAR(45) NULL,
     PRIMARY KEY (`id`));
   ```

How to do it...

> **Note**
>
> Since the last update from MySQL 8.0, we can't create users directly using the GRANT command. An error like this will appear:
>
> ```
> ERROR 1410 (42000): You are not allowed to create a user with GRANT
> ```
>
> To solve this issue, we will take some additional steps. We will also need at least two sessions of MySQL open, so keep that in mind if you opt to execute the commands directly on your command line.
>
> I would like to also thank Lefred's blog for this solution and contribution to the community. You can find more details and other useful information at their blog here: `https://lefred.be/content/how-to-grant-privileges-to-users-in-mysql-8-0/`.

Let us see the steps to perform this recipe:

1. First, let's create the `admin` user. Here, we will start to have problems if we do not follow the following steps correctly. We need to create a user to be our superuser or administrator, using our `root` user:

    ```
    CREATE user 'admin'@'localhost' identified by 'password';
    > Query OK, 0 rows affected
    ```

2. Then, we log in with the `admin` user. Using the password you defined in *step 1*, log in to the MySQL console using the `admin` user. We can't see any databases yet on the SQL software client, but we will fix this in the following step. After logging in to the console, you can see the SQL command ready to be used:

    ```
    $ mysql -u admin -p password -h localhost
    Type 'help;' or '\h' for help. Type '\c' to clear the current
    input statement.
    mysql>
    ```

    Keep this session open.

3. Then, we grant permissions to the `admin` user by role. In the root session, let's create a role called `administration` and grant full access to our database to the `admin` user:

    ```
    create ROLE administration;
    ```

    We then grant permissions to the role:

    ```
    grant
    alter,create,delete,drop,index,insert,select,update,trigger,
    alter routine,create routine, execute, create temporary tables
    on `cookbook_data`.* to 'administration';
    ```

    We then grant the role to our `admin` user:

    ```
    grant 'administration' to 'admin';
    ```

    We then set this role as the default:

    ```
    set default role 'administration' to 'admin';
    ```

4. Next, like in *step 2*, we create another two users, the `write` and `read-only` roles.

    Repeat *step 3*, giving the new roles names, and for these two roles, my recommendation is to grant the following privileges:

    *   **Writing permission**: `alter, create, index, insert, select, update, trigger, alter routine, create routine, execute,` and `create temporary tables`

    *   **Read-only permission**: `select` and `execute`

5. Next, we perform the actions. If we try to perform INSERT using our admin or write roles, we can see it is viable:

```
INSERT INTO `cookbook_data`.`people_city` (id, `name`, country,
occupation)
VALUES (1, 'Lin', 'PT', 'developer');
> 1 row(s) affected
```

However, the same can't be done by the read-only user:

```
Error Code: 1142. INSERT command denied to user
'reader'@'localhost' for table 'people_city'
```

How it works...

In a real-world project, most of the time, there is a dedicated person (or database administrator) to handle and take care of access to databases or data warehouses. Nonetheless, any person who needs to access a relational data source needs to understand the basic concepts of the access levels to ask for its permissions (and justify it).

In this recipe, we used a **Role-Based Access Control (RBAC)** model to define and attribute the role levels to the users. We created custom roles in our case, even though MySQL already has some built-in models. You can access the built-in models using the show privileges command.

alter, create, delete, drop, index, insert, select, update, trigger, alter routine, create routine, execute, and create temporary tables are the common commands used daily, and knowing them makes it easier to identify an error. Let's take, for example, the earlier error:

```
Error Code: 1142. INSERT command denied to user
'reader'@'localhost' for table 'people_city'
```

The first line shows precisely the permission we need. The second line shows us which user (reader) lacks permission, the connection they are using (@localhost), and the table (people_city) they want to access.

In addition, if you are a system administrator, you can also identify behavior that is not allowed and help to solve it.

There's more...

If you are interested to know more, it is also possible to find three other types of access control for databases and data warehouses:

- **Discretionary Access Control (DAC)**
- **Mandatory Access Control (MAC)**
- **Attribute-Based Access Control (ABAC)**

The following figure illustrates access control in a summarized version, as follows:

Attribute	RBAC	PBAC	ABAC
Network Access	Access to a network for a user is based on their role in an organization.	Access to a network is based on organizational policies as set by the administrator.	Access to a network for a user is based on user, environmental, or resource attributes.
Effectiveness	Effective if there is a clear role hierarchy that determines data access.	Effective for regulatory requirement demands and fits well with existing control schemes in IAM software like Azure AD.	Highly effective at defining data access.
Implementation	Simpler to implement as it aligns with organizational hierarchy.	Policies can be set once and used many times making implementation relatively simpler than ABAC.	Can become complex to sustain in the long run as it requires time and specialized effort.
Granular control	Lacks finer granular control.	Provides both coarse and fine control.	Provides for granular policy-based access using various attributes.

Figure 2.3 – A database access control comparison – source: https://www.cloudradius. com/access-control-paradigms-compared-rbac-vs-pbac-vs-abac/

Even though it is not described in the preceding figure, we can also find databases that use **row and column-based access control**. You can find out more about it here: `https://documentation. datavirtuality.com/23/reference-guide/authentication-access-control- and-security/access-control/data-roles/permissions/row-and-column- based-security`.

See also

- Here is an article by *Raimundas Matulevicius* and *Henri Lakk* on RBAC that discusses in depth the best approaches in several cases: `https://www.researchgate.net/publication/281479020_A_Model-driven_Role-based_Access_Control_for_SQL_Databases`

- This article by *Arundhati Singh* offers a perspective on RBAC mode implementation in enterprise networks: `https://dspace.mit.edu/bitstream/handle/1721.1/87870/53700676-MIT.pdf?sequence=2`

Accessing SSH File Transfer Protocol (SFTP) files

The **File Transfer Protocol** (**FTP**) was introduced in the 1970s at **Massachusetts Institute of Technology** (**MIT**) and is based on the **Transmission Control Protocol/Internet Protocol** (**TCP/IP**) application layer. Since the 1980s, it has been widely used to transfer files between computers.

Over the years, and with the increase in computer and internet usage, it became necessary to introduce a more secure way to use this solution. An **SSH layer** was implemented to improve the security of **FTP transactions**, creating the **SSH File Transfer Protocol** (**SFTP**) protocol.

Nowadays, it is common to ingest data from SFTP servers, and in this recipe, we will work to retrieve data from a public SFTP server.

Getting ready

In this recipe, we will create code with Python, using the `pysftp` library, to connect and retrieve sample data from a public SFTP server.

If you own an SFTP server, feel free to test the Python code here to exercise a little more:

1. First, we will get the SFTP credentials. Go to the SFTP.NET address at `https://www.sftp.net/public-online-sftp-servers`, and save the **Hostname** and **Login** (username/password) information on a notepad.

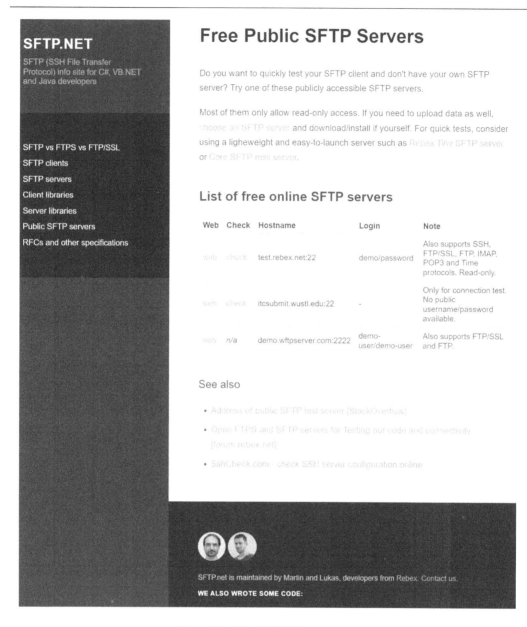

Figure 2.4 – The SFTP.NET main page

> **Note**
>
> This SFTP server is intended to be used only for testing and studying purposes. Because of that, the credentials are insecure and publicly available, and for production purposes, that information needs to be secured in a password vault and never shared through code.

2. Then, we install the Python `pysftp` package using the command line:

    ```
    $ pip install pysftp
    ```

How to do it...

Here are the steps to perform this recipe:

1. First, let's create a Python file called `accessing_sftp_files.py`. Then, we insert the following code to create our SFTP connection with Python:

    ```python
    import pysftp
    host = " test.rebex.net"
    username = "demo"
    password = "password"
    with pysftp.Connection(host=host, username=username,
    password=password) as sftp:
        print("Connection successfully established ... ")
    ```

    You can call the file using the following command:

    ```
    $ python accessing_sftp_files.py
    ```

    Here is the output for it:

    ```
    Connection successfully established ...
    ```

    Here, a known error might occur – `SSHException: No hostkey for host test.rebex.net found`.

    This happens because pysftp can't find hostkey in your KnowHosts.

    If this error occurs, follow the next steps:

2. Go to your command line and execute `ssh demo@test.rebex.net`:

    ```
    The authenticity of host 'test.rebex.net (195.144.107.198)' can't be established.
    ECDSA key fingerprint is SHA256:OzvpQxRUzSfV9F/ECMXbQ7B7zbK0aTngrhFCBUno65c.
    Are you sure you want to continue connecting (yes/no/[fingerprint])? yes
    Warning: Permanently added 'test.rebex.net,195.144.107.198' (ECDSA) to the list of known hosts.
    Password:
    ```

    Figure 2.5 – Adding a host to the known_hosts list

3. Insert the password for the demo user and exit Rebex Virtual Shell:

```
Welcome to Rebex Virtual Shell!
For a list of supported commands, type 'help'.
demo@ETNA:/$ exit
Disconnecting...
Connection to test.rebex.net closed.
```

Figure 2.6 – A welcome message from the Rebex SFTP server

4. Then, we list the files in the SFTP server:

```python
import pysftp

host = "test.rebex.net"
username = "demo"
password = "password"

with pysftp.Connection(host=host, username=username,
password=password) as sftp:
    print("Connection successfully established ... ")

    # Switch to a remote directory
    sftp.cwd('pub/example/')

    # Obtain structure of the remote directory '/pub/example'
    directory_structure = sftp.listdir_attr()

    # Print data
    for attr in directory_structure:
        print(attr.filename, attr)
```

Now, let's download the readme.txt file:

Let's change the last lines of our code to be able to download readme.txt:

```python
import pysftp

host = "test.rebex.net"
username = "demo"
password = "password"

with pysftp.Connection(host=host, username=username,
password=password) as sftp:
    print("Connection successfully established ... ")
```

```
# Switch to a remote directory
sftp.cwd('pub/example/')
print("Changing to pub/example directory... ")

sftp.get('readme.txt', 'readme.txt')
print("File downloaded ... ")
sftp.close()
```

The output for this is as follows:

```
Connection successfully established ...
Changing to pub/example directory...
File downloaded ...
```

How it works...

pysftp is a Python library that allows developers to connect, upload, and download data from SFTP servers. Its use is straightforward, and the library has tons of functionalities.

Note that most of our code is indented inside pysftp.Connection. This happens because we create a connection session for that particular credential. The with statement makes the acquisition and release of the resources, and as you can see, it is widely used in file streams, locks, sockets, and so on.

We also used the sftp.cwd() method, allowing us to change the directory and avoid specifying the path whenever we need to list or retrieve files.

Finally, the download was made using sftp.get(), where the first parameter is the path and the name of the file we want to download, and the second is where we will put it. Since we are already inside the file's directory, we can save it in our local HOME directory.

Last but not least, sftp.close() closes the connection. This is excellent practice in ingesting scripts to avoid network concurrency with other pipelines or the SFTP server.

There's more...

If you want to go deeper and make other tests, you can also create a local SFTP server.

For Linux users, it is possible to do this using the ssh command line. You can see more here: https://linuxhint.com/setup-sftp-server-ubuntu/.

For Windows users, go to the **SFTP Servers** section here: `https://www.sftp.net/servers`.

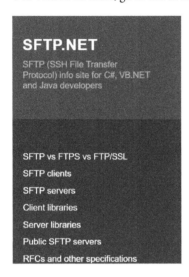

Figure 2.7 – The SFTP.NET page, with a link to a tutorial on how to create a small SFTP server

Select **Rebex Tiny SFTP Server** under **Minimalist SFTP servers**, download it, and start the program.

See also

- Creating a local SFTP server with Docker: `https://hub.docker.com/r/atmoz/sftp`

Retrieving data using API authentication

An **Application Programming Interface** (**API**) is a set of configurations that allows two systems or applications to communicate or transmit data with each other. Its concept has been improved in recent years, allowing faster transmissions and more security with **OAuth** methods, preventing **Denial of Service** (**DoS**) or **Distributed Denial of Service** (**DDoS**) attacks, and so on.

Its use is widely applied in data ingesting, whether to retrieve data from an application to retrieve the latest logs for analysis or from **BigQuery** using a cloud provider such as Google. Most applications nowadays make their data available through an API service, from which the data world gets a lot of benefits. The critical aspect here is to know how to retrieve data from an API service using the most accepted forms of authentication.

In this recipe, we will retrieve data from a public API using API key authentication, a standard method to gather data.

Getting ready

Since we will use two different methods, this section will be split to make it easy to understand how to handle them.

For this section, we will use the HolidayAPI, a public and free API that provides information about holidays worldwide:

1. Install the Python `requests` library:

    ```
    $ pip3 install requests
    ```

2. Then, access the Holiday API website. Go to `https://holidayapi.com/` and click on **Get Your Free API Key**. You should see the following page:

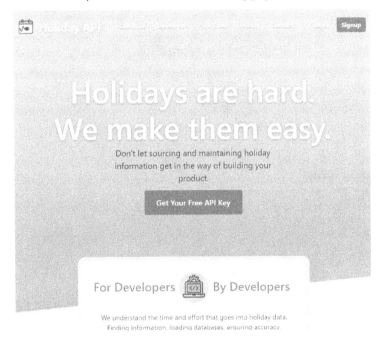

Figure 2.8 – The Holidays API main web page

3. Then, we create an account and get the API Key. To create an account, you can use an email and password or sign up with your GitHub account:

Figure 2.9 – The Holiday API user authentication page

After the authentication, you can see and copy your API Key. Note that you can also generate a new one at any time.

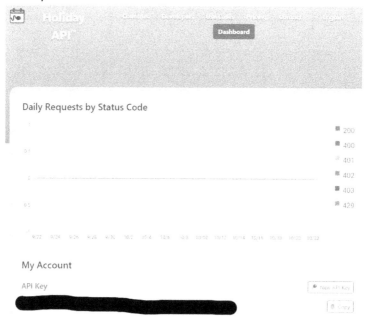

Figure 2.10 – The user dashboard page on the Holiday API page

> **Note**
>
> We will use the free tier of this API, which has limited requests per month. It is also prohibited from commercial use.

How to do it...

Here are the steps to perform the recipe:

1. We create a Python script using the `requests` library:

    ```python
    import requests
    import json

    params = { 'key': 'YOUR-API-KEY',
               'country': 'BR',
               'year': 2022
    }

    url = "https://holidayapi.com/v1/holidays?"

    req = requests.get(url, params=params)
    print(req.json())
    ```

    Ensure you use a `year` value equal to your previous year since we are using a free version of the API, which is limited to last year's historical data.

    Here is the output for the code:

    ```
    {'status': 200, 'warning': 'These results do not include
    state and province holidays. For more information, please
    visit https://holidayapi.com/docs', 'requests': {'used': 7,
    'available': 9993, 'resets': '2022-11-01 00:00:00'}, 'holidays':
    [{'name': "New Year's Day", 'date': '2021-01-01', 'observed':
    '2021-01-01', 'public': True, 'country': 'BR', 'uuid':
    'b58254f9-b38b-42c1-8b30-95a095798b0c',{...}
    ```

> **Note**
>
> As a best practice, the API key should never be hardcoded in the script. The definition here is for educational purposes.

2.   Then, we save our API request as a JSON file:

```
import requests
import json

params = { 'key': 'YOUR-API-KEY',
           'country': 'BR',
           'year': 2022
}

url = "https://holidayapi.com/v1/holidays?"
req = requests.get(url, params=params)

with open("holiday_brazil.json", "w") as f:
    json.dump(req.json(), f)
```

How it works...

The Python `requests` library is one of the most downloaded libraries on the PyPi servers. This popularity is not surprising, as we will see when we work with the library and see its power and versatility.

In *step 1*, we imported both the `requests` and `json` modules at the beginning of the Python script. The `params` dictionary is a payload sender to the API, so we inserted the API key and the two other mandatory fields.

> **Note**
>
> This API authorization key was sent through a payload request; however, it depends on how the API is built. Some request that the authentication is sent via `Header` definitions, for instance. Always check the API documentation or developer to understand how to authenticate correctly.

The `print()` function in *step 1* served as a test to see whether our calls were authenticated.

With the API call returning a `200` status code, we proceed to save the `JSON` file, and you should have output like this:

```
1    {
2      "status": 200,
3      "warning": "These results do not include state and province holidays. For more information, please visit https://holidayapi.com/docs",
4      "requests": {
5        "used": 9,
6        "available": 9991,
7        "resets": "2022-11-01 00:00:00"
8      },
9      "holidays": [
10       {
11         "name": "New Year's Day",
12         "date": "2021-01-01",
13         "observed": "2021-01-01",
14         "public": true,
15         "country": "BR",
16         "uuid": "b58254f9-b38b-42c1-8b30-95a095798b0c",
17         "weekday": {
18           "date": {
19             "name": "Friday",
20             "numeric": "5"
21           },
22           "observed": {
23             "name": "Friday",
24             "numeric": "5"
25           }
26         }
27       },
28       {
29         "name": "Shrove Monday",
30         "date": "2021-02-15",
31         "observed": "2021-02-15",
32         "public": false,
33         "country": "BR",
34         "uuid": "26346ac8-b1c7-4dfb-bda2-64c1ca445cc9",
35         "weekday": {
36           "date": {
37             "name": "Monday",
38             "numeric": "1"
39           },
40           "observed": {
41             "name": "Monday",
42             "numeric": "1"
43           }
44         }
45       },
46       {
```

Figure 2.11 – The downloaded JSON file data

There's more...

API keys are commonly used to authenticate clients, but other security methods such as OAuth should be considered, depending on the data sensitivity level.

> **Note**
> API key authentication can only be considered secure if associated with other security mechanisms such as HTTPS/SSL.

Authentication using the OAuth method

Open Authorization (**OAuth**) is an industry-standard protocol to authorize websites or applications to communicate and access information. You can find out more about it on the official documentation page here: `https://oauth.net/2/`.

You can also test this type of authentication with the **Google Calendar API**. To enable the OAuth method, follow these steps:

1. Enable the Google Calendar API by visiting the page at `https://developers.google.com/calendar/api/quickstart/python`, and then go to the **Enable APIs** section.

    A new tab will open, and **Google Cloud** will ask you to select or create a new project. Select the project you want to use.

> **Note**
>
> If you opt to create a new project, insert the project name in the **Project's Name** field and leave the **Organization** field with the default value (**No Organization**).

Select **Next** to confirm your project, and then click on **Activate**; you should then see this page:

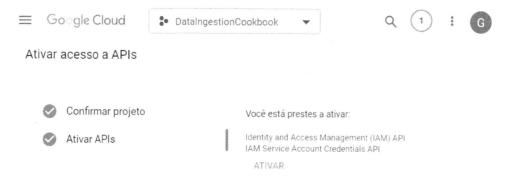

Figure 2.12 – The GCP page to activate a resource API

Now, we are almost ready to get our credentials.

2. Enable OAuth authentication by returning to the page at `https://developers.google.com/calendar/api/quickstart/python`, clicking on **Go to Credentials**, and following the instructions under **Authorize credentials for a desktop application**.

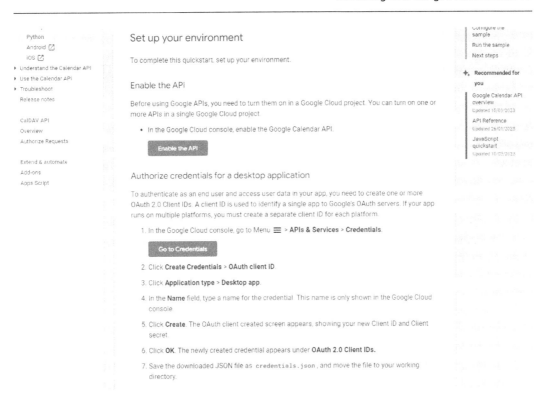

Figure 2.13 – GCP tutorial page to create the credentials.json file

At the end of this, you should have a `credentials.json` file for this recipe. Keep this file safe, since all calls to the Google API will require it to verify your authenticity.

You can use one of the GCP script examples to test this authentication method. Google has a Python script sample to retrieve data from the **Google Calendar API**, which can be accessed here: `https://github.com/googleworkspace/python-samples/blob/main/calendar/quickstart/quickstart.py`.

Other authentication methods

Even though we covered the two most common methods to authenticate to an API, a data ingestion pipeline is not limited to In your day-to-day work, you're likely to find legacy systems or applications that require other forms of authentication.

Methods such as **HTTP Basic and Bearer**, **OpenID Connect**, and **OpenAPI security schemes** are also widely used. You can find more details about them in this article written by *Guy Levin*: `https://blog.restcase.com/4-most-used-rest-api-authentication-methods/`.

SFTP versus API

You might be wondering, what is the difference between ingesting data from an SFTP server and an API? It is clear that they authenticate differently, and the code also behaves distinctly. But when should we implement an SFTP or API to make data available for ingestion?

FTP or SFTP transactions are designed to use **flat files**, such as **CSV**, **XML**, and **JSON** files. These two types of transactions also perform well when we need to transfer bulk data, and it is the only method available for older systems. An API provides real-time data delivery and a more secure internet-based connection, and its integration with several cloud applications has made it popular in the data ingestion world. However, API calls are sometimes paid for based on the number of requests.

This file transaction architecture discussion is primarily for financial and HR systems, which may use some old programming language versions. The system architecture discussion is based on flat files or real-time data in more recent and cloud-based applications.

See also

- You can find a list of public APIs with different types of auth methods here: `https://github.com/public-apis/public-apis`

- The Holiday API Python Client: `https://github.com/holidayapi/holidayapi-python`

- The Python `requests` library documentation: `https://requests.readthedocs.io/en/latest/user/quickstart/`

Managing encrypted files

When handling sensitive data is common, some fields or even the entire file is encrypted. It is comprehensive when this file security measure is implemented since sensitive data can expose the life of users. After all, encryption is the process of converting information into code that hides the original content.

Nonetheless, we must still ingest and process these encrypted files in our data pipelines. To be able to do so, we need to understand a bit more about how encryption works and how it is done.

In this recipe, we will decrypt a GnuPG-encrypted (where **GnuPG** stands for **GNU Privacy Guard**) file using Python libraries and best practices.

Getting ready

Before jumping into the fun part, we must install the GnuPG library on our local machine and download the encrypted dataset.

You will need two installations for the GnuPG file – one for the **operating system (OS)** and another for a Python package. This because the Python package requires internal resources from the installed OS package:

1. To use the Python wrapper library, we first need to install the GnuPG on our local machine:

    ```
    $ sudo apt-get install gnupg
    ```

    For Windows users, it is recommended to download the executable file here: `https://gnupg.org/download/index.html`.

    Mac users can install it using Homebrew: `https://formulae.brew.sh/formula/gnupg`.

2. Then, we install the Python GnuPG wrapper as follows:

    ```
    $ pip3 install python-gnupg
    Collecting python-gnupg
      Downloading python_gnupg-0.5.0-py2.py3-none-any.whl (18 kB)
    Installing collected packages: python-gnupg
    Successfully installed python-gnupg-0.5.0
    ```

3. Next, we download the **spotify tracks chart encrypted** dataset. You can use this link to download the file: `https://github.com/PacktPublishing/Data-Ingestion-with-Python-Cookbook/tree/main/Chapter_2/managing_encrypted_%EF%AC%81les`.

How to do it...

We will need a key to decrypt the file using GnuPG. You can find it in the **GitHub** repository of this book, inside the `Chapter 2 | Managing encrypted files` folder. The link to access it is in the *Technical requirements* section at the beginning of this chapter:

1. First, we import our key:

    ```
    import gnupg

    # Create gnupg directory
    gpg = gnupg.GPG(gnupghome='gpghome')

    # Open and import the key
    key_data = open('mykeyfile.asc').read()
    import_result = gpg.import_keys(key_data)

    # Show the fingerprint of our key
    print(import_result.results)
    ```

2. Then, we decrypt the ingestion file:

```
with open('spotify_data.csv.gpg', 'rb') as f:
    status = gpg.decrypt_file(f, passphrase='mypassphrase',
output='spotify_data.csv')

print(status.ok)
print(status.status)
print('error: ', status.stderr)
```

How it works...

Regarding the best practices for encrypting, GnuPG is a security reference and widely used, and it is documented in **RFC 4880**. You can find out more here: `https://www.rfc-editor.org/rfc/rfc4880`.

> **Note**
>
> A **Request for Comments** (**RFC**) is a technical documentation developed and maintained by the **Internet Engineering Task Force** (**IETF**). This institute specifies the best practices for protocols, services, and patterns on the internet.

Here, we saw a real-life example of GnuPG application, even though it seems simple. Let's pass through some important lines in the code:

In the `gpg = gnupg.GPG(gnupghome='gpghome')` line, we instantiated our GPG class and passed where it can store temporary files, and you can set any path you want to. In my case, I created a folder in my home directory called `gpghome`.

In the next lines, the key is imported, and we print its fingerprint just for demonstration purposes.

For *step 2*, we open the file we want to decrypt using the `with open` statement and decrypt it. You can see that a parameter for `passphrase` was set. This happened because the more recent versions of GnuPG require the file to have a passphrase set when encrypted. Since this recipe is only for educational matters, the passphrase here is simple and hardcoded.

After that, you should be able to open the `.csv` file with no problems.

	A	B	C	D	E
1	artist_name	track_name	track_id	popularity	
2	Juice WRLD	All Girls Are The Same	4VXIryQM	83	
3	Schoolgirl Byebye	Year,2015	0UsmyJDs	25	
4	Juice WRLD	Lucid Dreams	285pBltuF	84	
5	Fleetwood Mac	Rhiannon (Will You Ever Win) - 2018 Remaster	4fbwTO3C	49	
6	Joji	SLOW DANCING IN THE DARK	0rKtyWc8k	83	
7	Revenant	Year 2018	7AFlWfLvu	0	
8	Morgan Wallen	Whiskey Glasses	6foY66mV	78	
9	Frank Sinatra	It Was A Very Good Year - 2008 Remastered	1vLPTWPf.	43	
10	Lil Baby	Drip Too Hard (Lil Baby & Gunna)	78QR3Wp	80	
11	Fleetwood Mac	Gypsy - 2018 Remaster	5nTnApD6	46	
12	Billie Eilish	lovely (with Khalid)	0u2P5u6lv	87	
13	Anthem Lights	K-LOVE Fan Awards: Songs of the Year (2015 Mash-Up)	5rJw9VsPN	31	
14	girl in red	we fell in love in october	1BYZxKSf0	83	
15	Fleetwood Mac	Tusk - 2018 Remaster	6U0NPtEx.	48	
16	XXXTENTACION	SAD!	3ee8Jmje8	84	
17	Anthem Lights	K-Love Fan Awards: Songs of the Year (2014 Mash-Up)	00ohlpPn9	26	
18	Luke Combs	Beautiful Crazy	2rxQMGV.	78	
19	Fleetwood Mac	Dreams - 2018 Remaster	4pbO4Yljn	48	
20	Lil Baby	Yes Indeed	6vN77lE9L	78	
21	The Kiboomers	The Months of the Year - 2014 Version	58lQgf1Y5	28	
22	Playboi Carti	Shoota (feat. Lil Uzi Vert)	2BJSMvO(	78	
23	Fleetwood Mac	The Chain - Live 1997			
24	Travis Scott	SICKO MODE	2xLMifQCj	83	
25	Tori Amos	Pretty Good Year - 2015 Remaster	70euagGK	23	
26	Post Malone	Sunflower - Spider-Man: Into the Spider-Verse	3KkXRkHb	82	

Figure 2.14 – The decrypted Spotify CSV file

There's more…

Usually, GnuPG is the tool of choice when it comes to encrypting files, but there are other market solutions such as the Python `cryptography` library, which has a `Fernet` class, a symmetric encryption method. As you can see in the following code, its use is very similar to what we did in this recipe:

```
from cryptography.fernet import Fernet
# Retrieving key
fernet_key = Fernet(key)

# Getting and opening the encrypted file
with open('spotify_data.csv', 'rb') as enc_file:
    encrypted = enc_file.read()

# Decrypting the file
decrypted = fernet_key.decrypt(encrypted)

# Creating a decrypted file
with open('spotify_data.csv', 'wb') as dec_file:
    dec_file.write(decrypted)
```

Still, the `Fernet` method is not widely used in the data world. It happens because an encrypted file with sensitive data often comes from an application or software that uses GnuPG hybrid encryption, which, as we saw in the *How it works...* section, complies with RFC 4880.

You can find more details in the `cryptography` library documentation: `https://cryptography.io/en/latest/fernet/`.

See also

- How to create and encrypt files using the **Python Wrapper for GnuPG**: `https://gnupg.readthedocs.io/en/latest/`.

- GnuPG official page and documentation: `https://gnupg.org/`.

- If you are curious about RFC 4880 and want to understand it deeper, a summarized article about it was written by David Steele in his blog: `https://davesteele.github.io/gpg/2014/09/20/anatomy-of-a-gpg-key/`.

- *Voltage by opentext* is a great tool for data security and recommended by many companies. You can find out more here: `https://www.microfocus.com/en-us/cyberres/data-privacy-protection`.

Accessing data from AWS using S3

AWS is one of the most popular cloud providers, mixing different service architectures and allowing easy and fast implementations.

While it has various solutions for relational and non-relational databases, in this recipe, we will cover how to manage data access from **S3 buckets**, which is an object storage service allowing not only text files to be uploaded, but also media and several other types of files used in the IoT and big data fields.

There are two commonly used types of data access management for S3 buckets, both used on ingest pipelines – **user control** and **bucket policies**. In this recipe, we will learn how to manage access by user control, given that it is the most used method among data ingestion pipelines.

Getting ready

To do this recipe, having or creating an AWS account is not mandatory. The objective is to build a step-by-step **Identity Access Management** (**IAM**) policy to retrieve data from an S3 bucket using good data access practices you understand.

However, if you want to create a free AWS account to test it, you can follow the steps provided by the **AWS official docs** here: `https://docs.aws.amazon.com/accounts/latest/reference/manage-acct-creating.html`. After creating your AWS account, follow these steps:

1. Let's create a user to test our S3 policy. To create a user, check out this AWS link: `https://docs.aws.amazon.com/IAM/latest/UserGuide/id_users_create.html`. You don't need to worry about attaching policies to it, so skip this part of the tutorial. The idea is to explore what a user without any policies attached can do inside the AWS console.

2. Next, let's create an S3 bucket using our administrator user. On the search bar, type S3 and click on the first link:

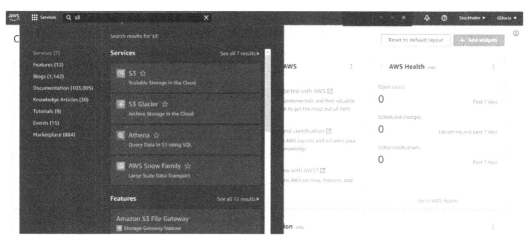

Figure 2.15 – The AWS search bar

3. Then, click the **Create bucket** button and a new page will load as follows:

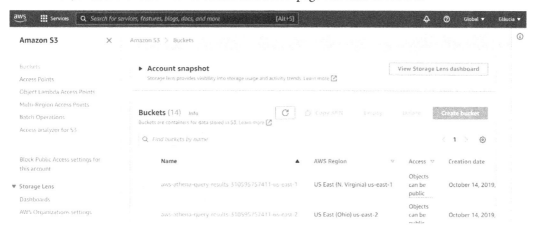

Figure 2.16 – The AWS S3 main page

A new page will load as follows:

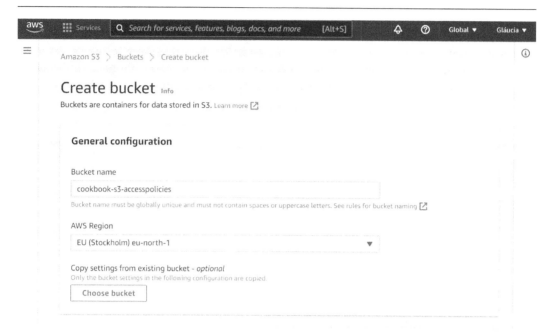

Figure 2.17 – The AWS S3 page to create a new bucket

On the **Create Bucket** page, insert the name of your bucket (it must be a unique name) and select the region you want to use. I will use `Stockholm` in this recipe since it is the nearest region to where I live. Skip the other fields for now, scroll down, and press the **Create bucket** button:

On the S3 page, you know should be able to see and select the bucket you created:

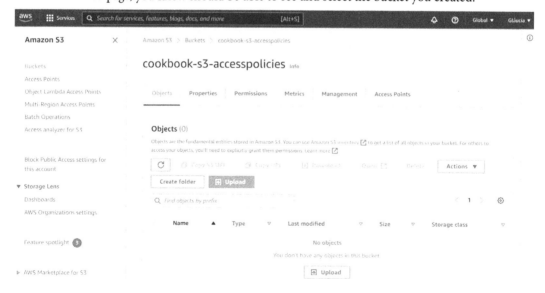

Figure 2.18 – The S3 bucket objects page

You can upload any files here for testing if you want to.

After completing the steps, open another window on your browser and switch to the user you created for testing. If possible, try to keep administrator and testing users logged on in different browsers so that you can see the changes in real time.

We are ready to start creating and applying the access policy.

How to do it...

If we try to list **S3**, we can see the buckets, but when clicking on any bucket, this error will happen:

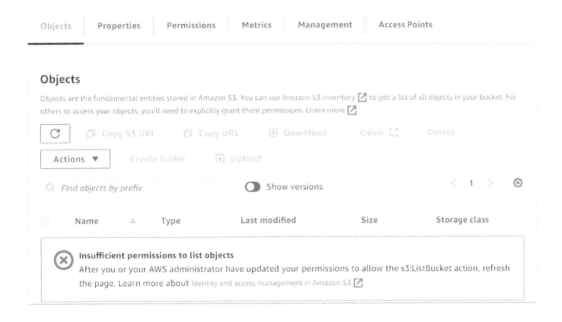

Figure 2.19 – Testing the user view of an S3 bucket with an Insufficient permissions message

Let's fix this by creating and attaching a policy with the following steps:

1. First, we will define an access policy for the user. The user will be capable of listing, retrieving, and deleting any object inside the S3 bucket we created. Let's start by creating a JSON file with the AWS policy requirements to make it possible. See the following file:

```
{
    "Version":"2012-10-17",
    "Statement":[
        {
            "Effect":"Allow",
            "Action":  "s3:ListAllMyBuckets",
            "Resource":"*"
        },
        {
            "Effect":"Allow",
            "Action":["s3:ListBucket","s3:GetBucketLocation"],
            "Resource":"arn:aws:s3:::cookbook-s3-accesspolicies"
        },
        {
            "Effect":"Allow",
            "Action":[
                "s3:PutObject",
                "s3:PutObjectAcl",
                "s3:GetObject",
                "s3:GetObjectAcl",
                "s3:DeleteObject"
            ],
            "Resource":"arn:aws:s3:::cookbook-s3-accesspolicies /*"
        }
    ]
}
```

2. Next, we allow the user to access the bucket via IAM policies. On the user IAM page, click on
    Add inline policy as follows:

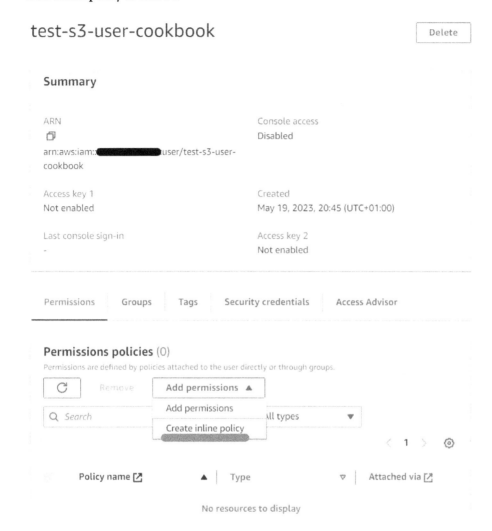

Figure 2.20 – AWS user Permission policy section

Insert the previous code into the **JSON** option tab, and click on **Review policy**. On the **Review Policy**
page, insert the policy's name and click **Create policy** to confirm it, as follows:

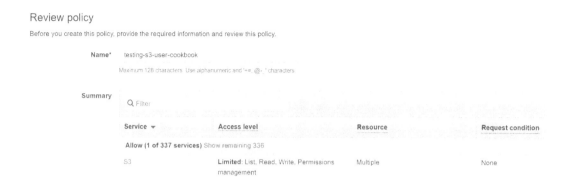

Figure 2.21 – The AWS IAM review policy page

If we check now, we can see that adding but not deleting files is possible.

How it works...

At the beginning of this recipe, our testing user had no permission to access any resources at AWS. For example, when accessing our created bucket, a warning appeared on the page. Then we allowed the user to access and upload files or objects to the bucket.

In *step 1*, we built an inline IAM policy with the following steps:

1.   First, we allowed the testing user to list all the buckets inside the respective AWS account:

```
"Statement":[
    {
        "Effect":"Allow",
        "Action": "s3:ListAllMyBuckets",
        "Resource":"*"
    },
```

2.   The second statement allows the user to list objects and get the bucket's location. Note that in the **Resource** key, we only specified a target S3 bucket **AWS Resource Name (ARN)**:

```
    {
        "Effect":"Allow",
        "Action":["s3:ListBucket","s3:GetBucketLocation"],
        "Resource":"arn:aws:s3:::cookbook-s3-accesspolicies"
    },
```

3. Finally, we create another statement to allow the insertion and retrieval of objects. In this case, the resource now also has a /* character at the end of the ARN. This represents the policy that is going to affect the respective bucket objects:

```
{
        "Effect":"Allow",
        "Action":[
            "s3:PutObject",
            "s3:PutObjectAcl",
            "s3:GetObject",
            "s3:GetObjectAcl",
        ],
        "Resource":"arn:aws:s3:::cookbook-s3-accesspolicies /*"
}
```

Depending on the AWS resource you want to manage access, the **Action** key can be very different and can have different applications. Regarding S3 buckets and objects, you find all possible actions in the AWS documentation: `https://docs.aws.amazon.com/AmazonS3/latest/userguide/using-with-s3-actions.html`.

There's more...

When ingesting data, the user control method is the most used. It happens because an application such use **Airflow** or **Elastic MapReduce** (**EMR**) can usually connect to a bucket. Also, from a management control perspective, it is much easier to handle, with just a few programmatic accesses instead of one for each user in a company.

Of course, there will be scenarios where each data engineer has a user with permissions set. Still, the scenario usually is (and should be) a development environment with a sample of data.

Bucket policies

Bucket policies can add a security layer to control the access of external resources to internal objects. With these policies, limiting access to specific IP addresses, specific resources such as **CloudFront**, or types of **HTTP** method requests is possible. In the AWS official documentation, you can see a list of practical examples: `https://docs.aws.amazon.com/AmazonS3/latest/userguide/example-bucket-policies.html`.

See also

In the AWS official documentation, you can also see other types of access control, such as **Access Control Lists** (**ACLs**) and **Cross-Origin Resource Sharing** (**CORS**): `https://docs.aws.amazon.com/AmazonS3/latest/userguide/s3-access-control.html`.

Accessing data from GCP using Cloud Storage

Google Cloud Platform (**GCP**) is a cloud provider that offers manifold services, from cloud computing to **Artificial Intelligence** (**AI**), which can be implemented in only a few steps. It also provides broad-spectrum storage called **Cloud Storage**.

In this recipe, we will build step-by-step policies to control access to data inside our **Cloud Storage buckets**.

Getting ready

This recipe will use the *uniform* method, as defined by the Google Cloud team:

1. First, we will create a testing user. Go to the **IAM** page (`https://console.cloud.google.com/iam-admin/iam`) and select **Grant Access**. Add a valid Gmail address in the **New principals** field. For now, this user will only have the **Browser** role:

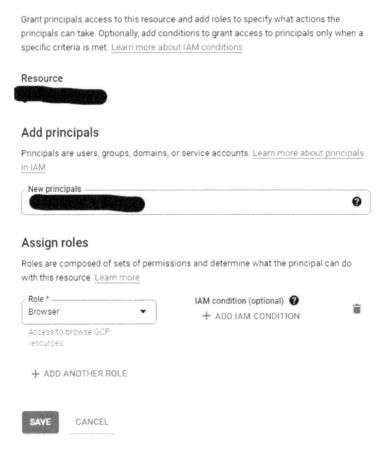

Figure 2.22 – The GCP IAM page to attach policies to a user

2. Then, we will create a Cloud Storage bucket. Go to the **Cloud Storage** page and select **Create a bucket**: `https://console.cloud.google.com/storage/create-bucket`.

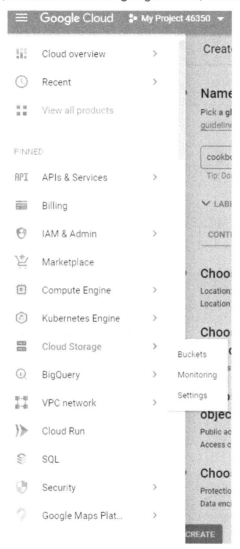

Figure 2.23 – The GCP search bar with Cloud Storage selected

Add a unique name to your bucket and leave the other option as it is:

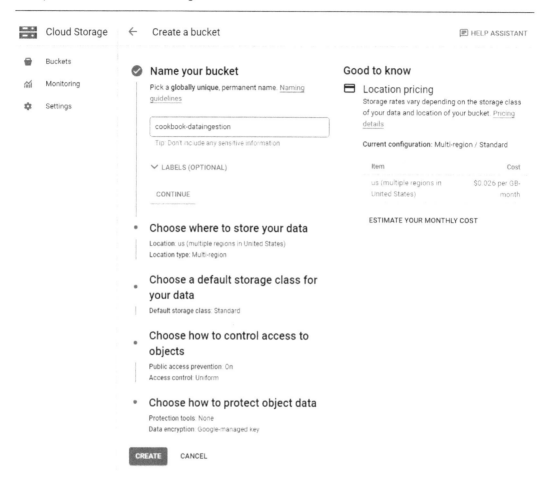

Figure 2.24 – The GCP page to create a new bucket

How to do it...

Here are the steps to perform this recipe:

1. We will try to access the Cloud Storage objects. First, let's try to access the bucket using the user we just created. An error message should appear:

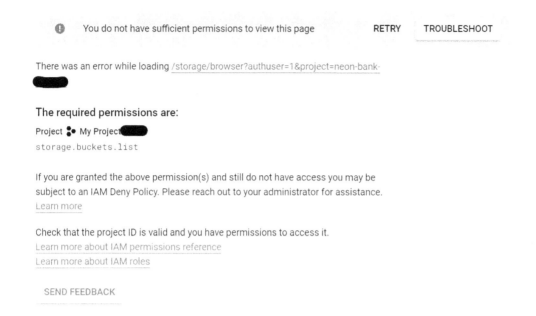

Figure 2.25 – An insufficient permission message for the testing user in the GCP console

2.   Then, we grant **Editor** permissions in Cloud Storage. Go to the **IAM** page and select the testing user you created. On the editing user page, select **Editor**.

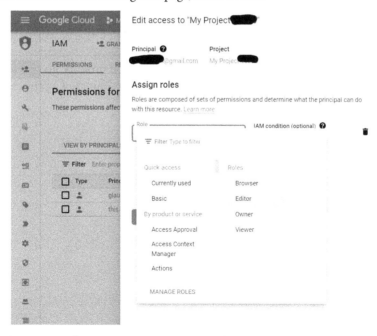

Figure 2.26 – The GCP IAM page – assigning the Editor role to the testing user

> **Note**
>
> If you are confused about the roles, use the Policy Simulator in Google Cloud: `https://console.cloud.google.com/iam-admin/simulator`.

3. Next, we access Cloud Storage with a proper role. The user should be able to see and upload objects to the bucket:

Figure 2.27 – The testing user view of the GCP bucket

How it works...

In *step 1*, our testing user had only permission to browse, and an error message appeared when trying to see the bucket list. However, the **Editor** role for Cloud Storage solves this problem by granting access to the bucket (and most of the other essential Google Cloud resources). At this point, creating a condition to allow only access to this bucket is also possible.

The Google Cloud access hierarchy is based on its organization and projects. To provide access to a respective bucket, we need to ensure we also have access to the resources of a project. Refer to the following screenshot:

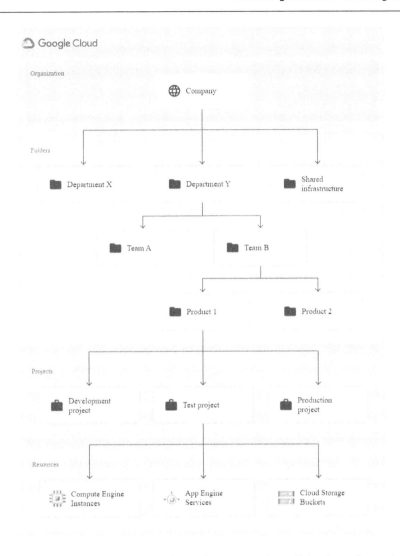

Figure 2.28 – GCP control access hierarchy – source: https://cloud.google.com/
resource-manager/docs/cloud-platform-resource-hierarchy#inheritance

Once the access hierarchy is defined, we can select several built-in user permission groups on the IAM page and add conditions if needed. Unlike AWS, Google Cloud policies are often created as "roles" and grouped to serve a specific area or department. Additional permissions or conditionals can be made for particular cases, but they are not shared.

Even though the uniform method seems simple, it can be a powerful way to manage access to Google Cloud when properly grouped and revised and uniformly grant permissions. In our case, the **Editor** role solved our problem, but a specialist on this topic is recommended when working with larger teams and different types of access policies.

There's more...

Like S3, Cloud Storage also has another type of access control called **fine-grained**. It consists of a mixture of IAM policies and ACLs and is recommended in cases where storage connects to an S3 bucket, for instance. As its name suggests, the permission is refined to the bucket and individual object levels. It needs to be configured by someone (or a team) with a high level of security knowledge, since the data exposure can be elevated if the **ACL policy** is not set correctly.

You can read more about ACLs in Cloud Storage here: `https://cloud.google.com/storage/docs/access-control/lists`.

Further reading

- `https://www.manageengine.com/device-control/data-replication.html`
- `https://www.keboola.com/blog/database-replication-techniques`
- `https://satoricyber.com/access-control/access-control-101-a-comprehensive-guide-to-database-access-control/`
- `https://stackoverflow.com/questions/190257/best-role-based-access-control-rbac-database-model`
- `https://www.researchgate.net/publication/281479020_A_Model-driven_Role-based_Access_Control_for_SQL_Databases`
- `https://dspace.mit.edu/bitstream/handle/1721.1/87870/53700676-MIT.pdf?sequence=2`
- `https://scholarworks.calstate.edu/downloads/sb397840v`
- `https://www.sftp.net/public-online-sftp-servers`
- `https://www.ittsystems.com/how-to-access-sftp-server-in-python/`
- `https://towardsdatascience.com/encrypt-and-decrypt-files-using-python-python-programming-pyshark-a67774bbf9f4`
- `https://www.geeksforgeeks.org/encrypt-and-decrypt-files-using-python/`
- `https://www.saltycrane.com/blog/2011/10/python-gnupg-gpg-example/`
- `https://www.ekransystem.com/en/blog/data-security-best-practices`
- `https://www.ovaledge.com/blog/data-access-management-basics-implementation-strategy`

Data Discovery – Understanding Our Data before Ingesting It

As you may already have noticed, **data ingestion** is not just retrieving data from a source and inserting it in another place. It involves understanding some business concepts, secure access to the data, and how to store it, and now it is essential to discover our data.

Data discovery is the process of understanding our data's patterns and behaviors, ensuring the whole data pipeline will be successful. In this process, we will understand how our data is modeled and used, so we can set up and plan our ingestion using the best fit.

In this chapter, you will learn about the following:

- Documenting the data discovery process
- Configuring OpenMetadata
- Connecting OpenMetadata to our database

Technical requirements

You can also find the code from this chapter in its GitHub repository here: `https://github.com/PacktPublishing/Data-Ingestion-with-Python-Cookbook`.

Documenting the data discovery process

In recent years, manual data discovery has been rapidly deprecated, giving rise to **machine learning** and other automated solutions, bringing fast insights into data in storage or online spreadsheets, such as Google Sheets.

Nevertheless, many small companies are just starting out their businesses or data areas, so implementing a paid or cost-related solution might not be a good idea right away. As data professionals, we also need to be malleable when applying the first solution to a problem – there will always be space to improve it later.

Getting ready

This recipe will cover the steps to start the data discovery process effectively. Even though, here, the process is more related to the manual discovery steps, you will see it also applies to the automated ones.

Let's start by downloading the datasets.

For this recipe, we are going to use the *The evolution of genes in viruses and bacteria* dataset (`https://www.kaggle.com/datasets/thedevastator/the-evolution-of-genes-in-viruses-and-bacteria`), and another one containing *hospital administration* information (`https://www.kaggle.com/datasets/girishvutukuri/hospital-administration`).

> **Note**
>
> This recipe does not require the use of the exact datasets mentioned – it covers generically how to apply the methodology to datasets or any data sources. Feel free to use any data you want.

The next stage is creating the documentation. You can use any software or online application that suits you – the important thing is to have a place to detail and catalog the information.

I will use **Notion** (`https://www.notion.so/`). Its home page is shown in *Figure 3.1*. It offers a free plan and allows you to create separate places for different types of documentation. However, some companies use **Confluence by Atlassian** to document their data. It will always depend on the scenario you are in.

Figure 3.1 – Notion home page

This is an optional stage where we are creating a Notion account. On the main page, click on **Get Notion free**.

Another page will appear and you can use your Google or Apple email to create an account, as follows:

 Notion

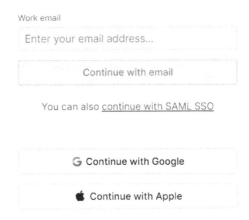

Figure 3.2 – Notion Sign up page

After that, you should see a blank page with a welcome message from Notion. If any other action is required, just follow the page instructions.

How to do it...

Let's imagine a scenario where we work at a hospital and need to apply the data discovery process. Here is how we go about it:

1. **Identifying our data sources**: Two main departments need their data to be ingested—the administration and research departments. We know they usually keep their CSV files in a local data center so we can access them via the intranet. Don't mind the filenames; generally, in a real application, they are not supported.

The following are the research department's files:

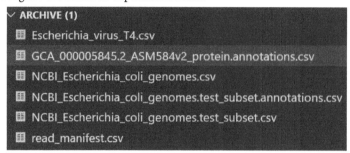

Figure 3.3 – Research files on the evolution of genes in E. coli

The following are the administration department's files:

Figure 3.4 – Hospital administration files

2. **Categorizing data per department or project**: Here, we create folders and subfolders related to the department and the type of data (on patients or specific diseases).

Figure 3.5 – Research Department page

3. **Identifying the datasets or databases**: When looking at the files, we can find four patterns. There are the exclusive datasets: **E.Coli Genomes**, **Protein Annotations**, **Escherichia Virus** in general, and **Patients**.

Figure 3.6 – Subsections created by research type and hospital administration topic

4. **Describing our data**: Now, at the dataset level, we need to have helpful information about it, such as the overall description of that dataset table, when it is updated, where other teams can find it, a description of each column of the table, and, last but not least, all metadata.

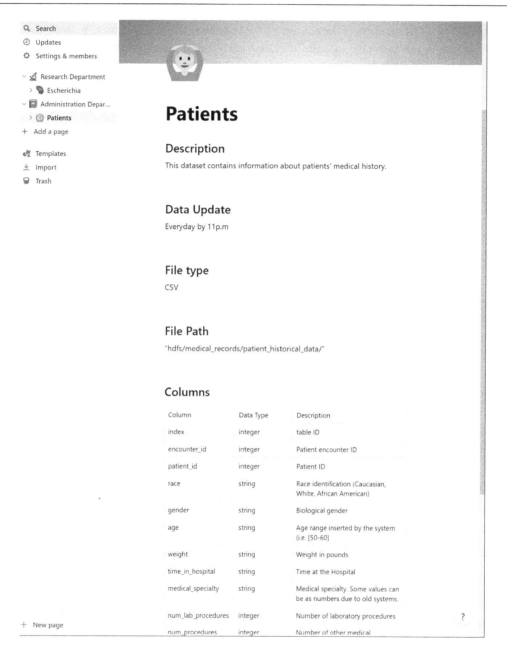

Figure 3.7 – Patient data documentation using Notion

Note

The description of where the file is stored may not be applied in all cases. You can find the reference of the database name instead, such as `'admin_database.patients'`.

How it works...

When starting data discovery, the first objective is identifying patterns and categorizing them to create a logical flow. Usually, the first categorizations are by department or project, followed by database and dataset identification, and finally, describing the data inside.

There are some ways to document data discovery manually. People more used to the old-fashioned style of **BI** (short for **Business Intelligence**) tend to create more beautiful visualization models to apply discovery. However, this recipe's objective is to create a catalog using a simple tool such as Notion:

1. **Categorizing data as per department or project**: The first thing we did was to identify the department responsible for each piece of data. Who is the contact in the case of an ingestion problem or if the dataset is broken? In formal terms, they are also known as data stewards. In some companies, categorization by project can also be applied since some companies can have their particular necessities and data.

2. **Identifying the datasets or databases**: Here, we have only used datasets. Under the projects and/or departments, we insert the name of each table and other helpful information. If the tables are periodically updated, it is a good practice to also document that.

3. **Describing our data**: Finally, we document the expected columns with their data types in detail. It helps data engineers plan their scripts when ingesting raw data; if something goes wrong after the automation, they can easily detect the issue.

You might notice that some data behaves strangely. For instance, the **medical_speciality** column in *Figure 3.7* has values described and a number to reference something else. In a real-world project, it would be necessary to create auxiliary data inside our ingestion to make a pattern and later facilitate the report or dashboards.

Configuring OpenMetadata

OpenMetadata is an open source tool used for metadata management, allowing the process of **data discovery** and **governance**. You can find more about it here: `https://open-metadata.org/`.

By performing a few steps, it is possible to create a local or production instance using **Docker** or **Kubernetes**. OpenMetadata can connect to multiple resources, such as **MySQL**, **Redis**, **Redshift**, **BigQuery**, and others, to bring the information needed to build a data catalog.

Getting ready

Before starting our configuration, we must install **OpenMetadata** and ensure the Docker containers are running correctly. Let us see how it is done:

> **Note**
>
> At the time this book was written, the application was in the 0.12 version and with some documentation and installation improvements. This means the best approach to installing it may change over time. Please refer to the official documentation for it here: `https://docs.open-metadata.org/quick-start/local-deployment`.

1. Let's create a folder and `virtualenv` (optional):

    ```
    $ mkdir openmetadata-docker
    $ cd openmetadata-docker
    ```

    Since we are using a Docker environment to deploy the application locally, you can create it with `virtualenv` or not:

    ```
    $ python3 -m venv openmetadata
    $ source openmetadata /bin/activate
    ```

2. Next, we install OpenMetadata as follows:

    ```
    $ pip3 install --upgrade "openmetadata-ingestion[docker]"
    ```

3. Then we check the installation, as follows:

    ```
    $ metadata
    Usage: metadata [OPTIONS] COMMAND [ARGS]...

      Method to set logger information

    Options:
      --version                         Show the version and exit.
      --debug / --no-debug
      -l, --log-level [INFO|DEBUG|WARNING|ERROR|CRITICAL]
                                        Log level
      --help                            Show this message and exit.

    Commands:
      backup                            Run a backup for the metadata
    DB.
      check
      docker                            Checks Docker Memory
    Allocation Run...
      ingest                            Main command for ingesting
    metadata...
      openmetadata-imports-migration    Update DAG files generated
    after...
      profile                           Main command for profiling
    Table...
    ```

```
   restore                    Run a restore for the metadata
DB.
   test                       Main command for running test
suites
   webhook                    Simple Webserver to test
webhook...
```

How to do it...

After downloading the **Python** package and **Docker**, we will proceed with the configurations as follows:

1. **Running containers**: It may take some time to finish when you execute it for the first time:

   ```
   $ metadata docker -start
   ```

> **Note**
>
> It is common for this type of error to appear:
>
> ```
> Error response from daemon: driver failed programming
> external connectivity on endpoint openmetadata_ingestion
> (3670b9566add98a3e79cd9a252d2d0d377dac627b4be94b669482f6ccce350e0):
> Bind for 0.0.0.0:8080 failed: port is already allocated
> ```
>
> It means other containers or applications are already using port 8080. To solve this, specify another port (such as 8081) or stop the other applications.

The first time you run this command, the results might take a while due to other containers associated with it.

In the end, you should see the following output:

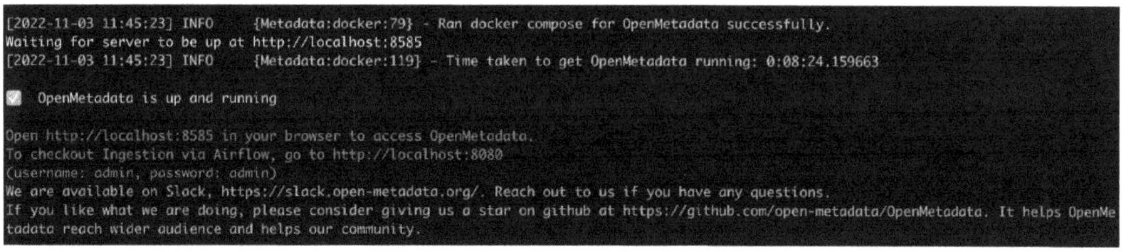

Figure 3.8 – Command line showing success running OpenMetadata containers

2. **Accessing OpenMetadata UI**: When the container installation is finished, you should be able to access the UI via a browser using the http://localhost:8585 address:

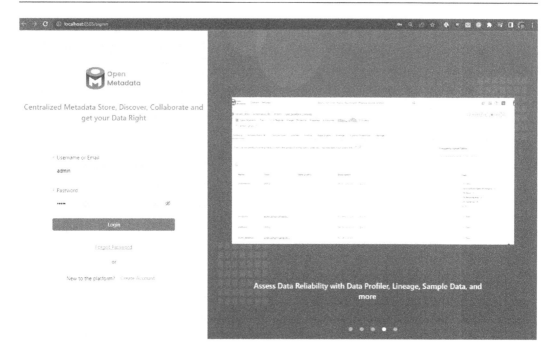

Figure 3.9 – OpenMetadata sign-in page in the browser

3. **Creating a user account and logging in**: To access the UI panel, we need to create a user account as follows:

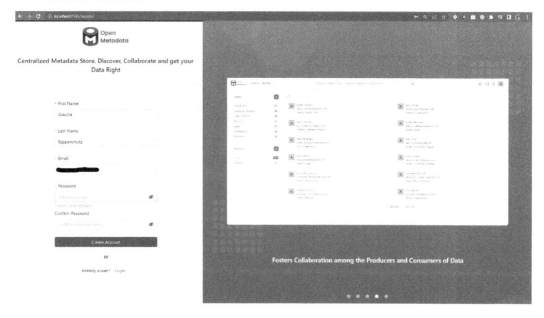

Figure 3.10 – Creating a user account in the OpenMetadata Create Account section

After that, we will be redirected to the main page and be able to access the panel, shown as follows:

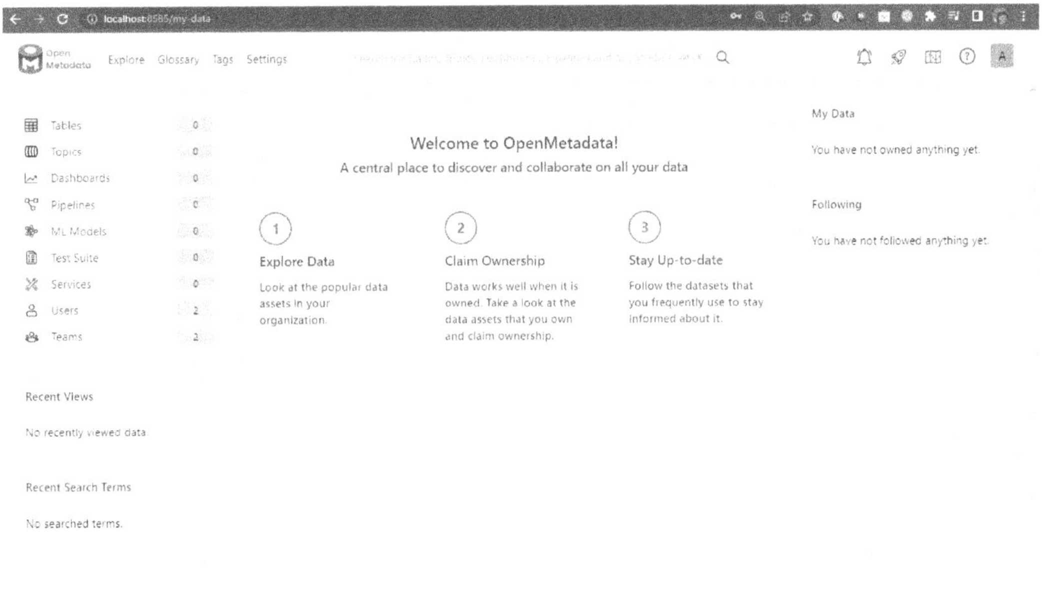

Figure 3.11 – Main page of OpenMetadata

> **Note**
>
> Is also possible to log in using the default admin user by inserting the admin@openmetadata.org username and admin as the password.
>
> For production matters, please refer to the Enable Security Guide here: https://docs.open-metadata.org/deployment/docker/security.

4. **Creating teams**: In the **Settings** section, you should see several possible configurations, from creating users to access the console to integrations with messengers such as **Slack** or **MS Teams**.

Some ingestion and integration requires the user to be allocated to a team. To create a team, we first need to log in as `admin`. Then, go to **Settings** | **Teams** | **Create new team**:

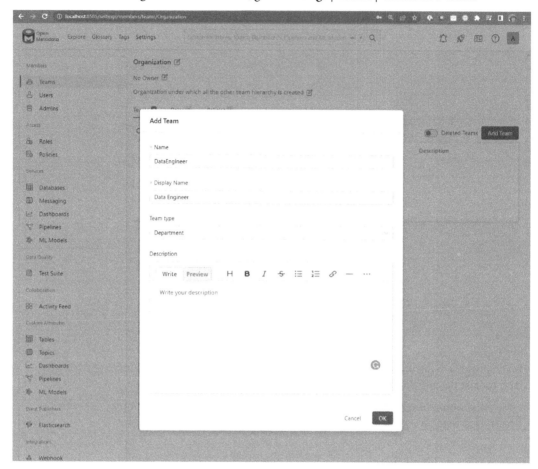

Figure 3.12 – Creating a team in the OpenMetadata settings

5. **Adding users to our teams**: Select the team you just created and go to the **Users** tab. Then select the user you want to add.

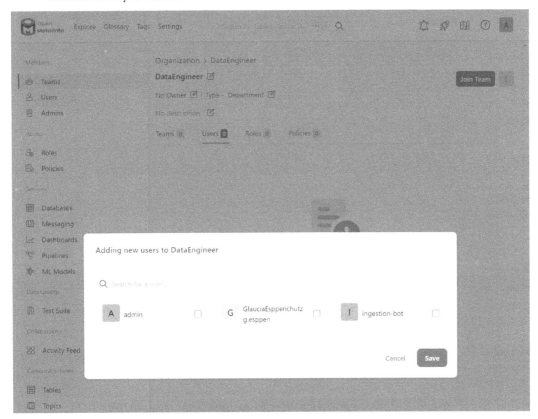

Figure 3.13 – Adding users to a team

Creating teams is very convenient to keep track of users' activity and define a group of roles and policies. In the following case, all users added to this team will be able to navigate through and create their data discovery pipelines.

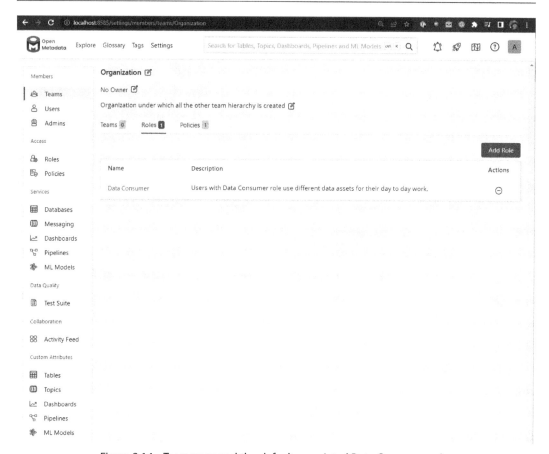

Figure 3.14 – Team page and the default associated Data Consumer role

We must have a Data Steward or Administrator role for the activities in this chapter and the following recipe. The Data Steward role has almost the same permissions as the Administrator role since it is a position that is responsible for defining and implementing data policies, standards, and procedures to govern data usage and ensure consistency.

You can read more about the **Roles and Policies** of OpenMetadata here: `https://github.com/ open-metadata/OpenMetadata/issues/4199`.

How it works...

Now, let's understand a bit more about how OpenMetadata works.

OpenMetadata is an open source metadata management tool designed to help organizations to manage their data and metadata across different systems or platforms. Since it centralizes data information in one place, it makes it easier to discover and understand data.

It is also a flexible and extensible tool, allowing integration with tools such as Apache Kafka, Apache Hive, and others since it uses programming languages such as **Python** (main core code) and Java behind the scenes.

To orchestrate and ingest the metadata from sources, OpenMetadata counts the sources using Airflow code. If you look at its core, all Airflow code can be found in `openmetadata-ingestion`. For more heavy users who want to debug any problems related to the ingestion process in this framework, Airflow can be easily accessed at `http://localhost:8080/`, when the metadata Docker container is up and running.

It also uses **MySQL DB** to store user information and relationships and an **Elasticsearch** container to create efficient indexes. Refer to the following figure (`https://docs.open-metadata.org/developers/architecture`):

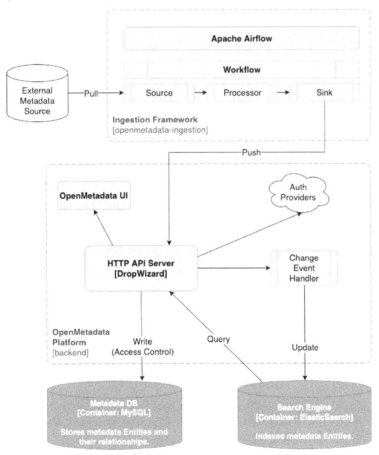

Figure 3.15 – OpenMetadata architecture diagram

Font source: OpenMetadata documentation

For more detailed information about the design decisions, you can access the **Main Concepts** page and explore in detail the ideas behind them: `https://docs.open-metadata.org/main-concepts/high-level-design`.

There's more...

We saw how **OpenMetadata** can be easily configured and installed locally on our machines and a brief overview of its architecture. However, other good options on the market can be used to document data, or even a **SaaS** solution of **OpenMetadata** using **Google Cloud**.

OpenMetadata SaaS sandbox

Recently, OpenMetadata implemented a **Software as a Service** (**SaaS**) sandbox (`https://sandbox.open-metadata.org/signin`) using Google, making it easier to deploy and start the discovery and catalog process. However, it may have costs applied, so keep that in mind.

See also

- You can read more about OpenMetadata in their blog: `https://blog.open-metadata.org/why-openmetadata-is-the-right-choice-for-you-59e329163cac`

- Explore OpenMetadata on GitHub: `https://github.com/open-metadata/OpenMetadata`

Connecting OpenMetadata to our database

Now that we have configured our **Data Discovery** tool, let's create a sample connection to our local database instance. Let's try to use PostgreSQL to do an easy integration and practice another database usage.

Getting ready

First, ensure our application runs appropriately by accessing the `http://localhost:8585/my-data` address.

> **Note**
> Inside OpenMetadata, the user must have the **Data Steward** or **Administration** role to create connections. You can switch to the `admin` user using the previous credentials we saw.

You can check the Docker status here:

Figure 3.16 – Active containers are shown in the Docker desktop application

Use PostgreSQL for testing. Since we already have a Google project ready, let us create a SQL instance using the PostgreSQL engine.

As we kept the queries to create the database and tables in *Chapter 2*, we can build it again in Postgres. The queries can also be found in the GitHub repository of this chapter. However, feel free to create your own data.

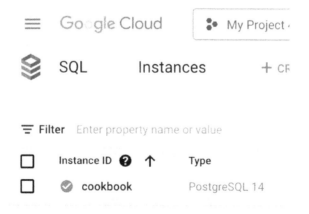

Figure 3.17 – Google Cloud console header for SQL instances

Remember to let this instance allow public access; otherwise, our local OpenMetadata instance won't be able to access it.

How to do it...

Go to the OpenMetadata home page by typing `http://localhost:8585/my-data` in the browser header:

1. **Adding a new database to OpenMetadata**: Go to **Settings | Services | Databases** and click on **Add new Database Service**. Some options will appear. Click on **Postgres**:

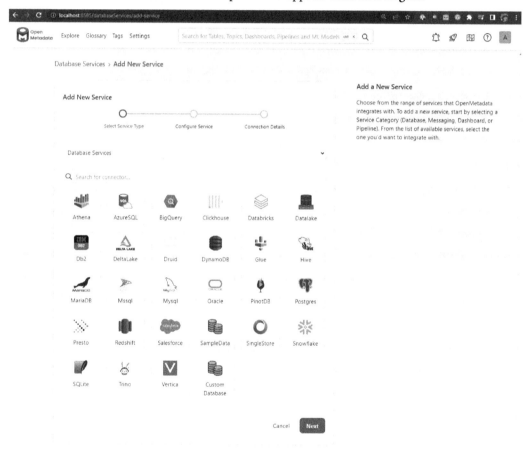

Figure 3.18 – OpenMetadata page to add a database as a source

Click on **Next** and add a service name. It can be anything you like since it's an identifier. I used CookBookData.

2. **Adding our connection settings**: After clicking on **Next** again, a page with some fields to input the MySQL connection settings will appear:

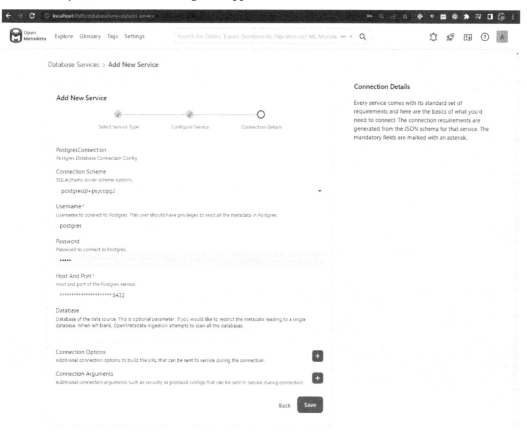

Figure 3.19 – Adding new database connection information

3. **Testing our connection**: With all the credentials in place, we need to test the connection to the database.

Database

Database of the data source. This is optional parameter, if you would like to restrict the metadata reading to a single database. When left blank, OpenMetadata Ingestion attempts to scan all the databases.

Connection Options

Additional connection options to build the URL that can be sent to service during the connection.

Connection Arguments

Additional connection arguments such as security or protocol configs that can be sent to service during connection.

OpenMetadata will connect to your resource from the IP 188.83.27.166. Make sure to allow inbound traffic in your network security settings.

Connection test was successful   Test Connection

Back   Save

Figure 3.20 – Connection test successful message for database connection

4. **Creating an ingestion pipeline**: You can leave all the fields as they are without worrying about the **database tool (DBT)**. For **Schedule Interval**, you can set what suits you best. I will leave it as **Daily**.

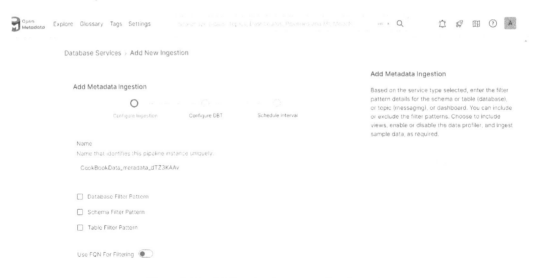

Figure 3.21 – Adding database metadata ingestion

5. **Ingesting the metadata**: Heading to **Ingestions**, our database metadata is successfully ingested.

Figure 3.22 – Postgres metadata successfully ingested

6. **Exploring our metadata**: To explore the metadata, go to **Explore | Tables**:

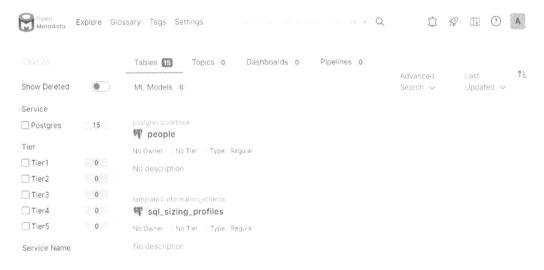

Figure 3.23 – Explore page showing the tables metadata ingested

You can see that the `people` table is there with other internal tables:

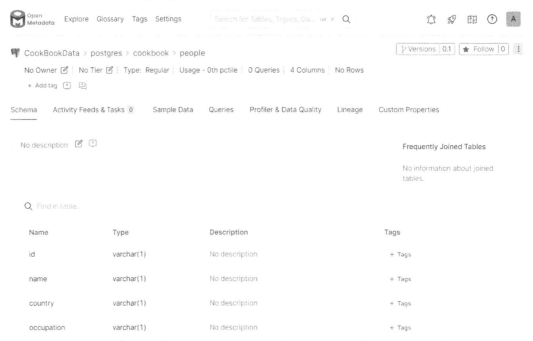

Figure 3.24 – The people table metadata

Here, you can explore some functionalities of the application, such as defining the level of importance to the organization and the owners, querying the table, and others.

How it works...

As we saw previously, OpenMetadata uses Python to build and connect to different sources.

In **Connection Details**, we saw `Connection Scheme` uses `psycopg2`, a widely used library in Python. All other arguments are passed to the behind-the-scenes Python code to create a connection string.

For each metadata ingestion, OpenMetadata will create a new Airflow **Directed Acyclic Graph** (**DAG**) to process it based on a generic one. Having a separate DAG for each metadata ingestion makes debugging more manageable in case of errors.

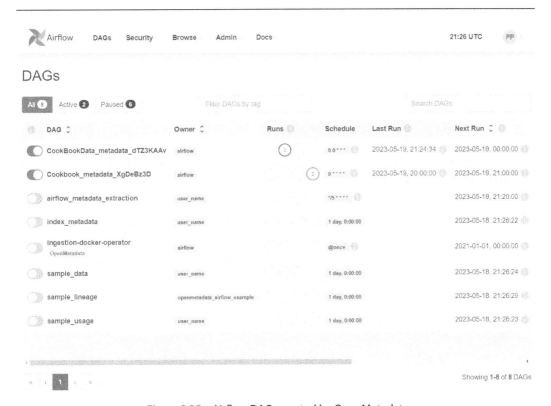

Figure 3.25 – Airflow DAGs created by OpenMetadata

If you open the Airflow instance used by OpenMetadata, you can see it clearly and have other information about the metadata ingestion. It's a nice place to debug in case an error occurs. Understanding how our solution works and where to look in case of a problem helps identify and solve issues more efficiently.

Further reading

- https://nira.com/data-discovery/
- https://coresignal.com/blog/data-discovery/
- https://www.polymerhq.io/blog/diligence/what-is-data-discovery-guide/
- https://bi-survey.com/data-discovery
- https://www.heavy.ai/technical-glossary/data-discovery
- https://www.datapine.com/blog/what-are-data-discovery-tools/
- https://www.knowsolution.com.br/data-discovery-como-relaciona-bi-descubra/ (in Portuguese)

Other tools

If you are interested in learning more about other data discovery tools available on the market, here are some:

- **Tableau**: Tableau (`https://www.tableau.com/`) is more extensively used for data visualizations and dashboards but comes with some features to discover and catalog data. You can read more about how to use Tableau for data discovery on their resources page here: `https://www.tableau.com/learn/whitepapers/data-driven-organization-7-keys-data-discovery`.

- **OpenDataDiscovery** (free and open source): OpenDataDiscovery has recently arrived on the market and can provide a very nice starting point. Check it out here: `https://opendatadiscovery.org/`.

- **Atlan**: Atlan (`https://atlan.com/`) is a complete solution and also brings a data governance structure; however, the costs can be high and it requires a call with their sales team to start an **MVP** (short for **Minimum Viable Product**).

- **Alation**: Alation is an enterprise tool that provides several data solutions that include all pillars of data governance. Find out more here: `https://www.alation.com/`.

4

Reading CSV and JSON Files and Solving Problems

When working with data, we come across several different types of data, such as structured, semi-structured, and non-structured, and some specifics from other systems' outputs. Yet two widespread file types are ingested, **comma-separated values (CSV)** and **JavaScript Object Notation (JSON)**. There are many applications for these two files, which are widely used for data ingestion due to their versatility.

In this chapter, you will learn more about these file formats and how to ingest them using Python and PySpark, apply the best practices, and solve ingestion and transformation-related problems.

In this chapter, we will cover the following recipes:

- Reading a CSV file
- Reading a JSON file
- Creating a SparkSession for PySpark
- Using PySpark to read CSV files
- Using PySpark to read JSON files

Technical requirements

You can find the code for this chapter in this **GitHub** repository: `https://github.com/PacktPublishing/Data-Ingestion-with-Python-Cookbook`.

Using **Jupyter Notebook** is not mandatory but can help you see how the code works interactively. Since we will execute Python and PySpark code, it can help us understand the scripts better. Once you have installed it, you can execute Jupyter using the following command:

```
$ jupyter notebook
```

It is recommended to create a separate folder to store the Python files or notebooks we will create in this chapter; however, feel free to organize them however suits you best.

Reading a CSV file

A **CSV** file is a plain text file where commas separate each data point, and each line represents a new record. It is widely used in many areas, such as finance, marketing, and sales, to store data. Software such as **Microsoft Excel** and **LibreOffice**, and even online solutions such as **Google Spreadsheets**, provide reading and writing operations for this file. Visually it resembles a structured table, which greatly enhances the file's usability.

Getting ready

You can download the CSV dataset for this from **Kaggle**. Use this link to download the file: `https://www.kaggle.com/datasets/jfreyberg/spotify-chart-data`. We are going to use the same Spotify dataset as in *Chapter 2*.

> **Note**
> Since Kaggle is a dynamic platform, the filename might change occasionally. After downloading it, I named the file `spotify_data.csv`.

For this recipe, we will use only Python and Jupyter Notebook to execute the code and create a more friendly visualization.

How to do it...

Follow these steps to try this recipe:

1. We begin by reading the CSV file:

    ```
    import csv

    filename = "     path/to/spotify_data.csv"
    columns = []
    rows = []
    ```

```
with open (filename, 'r', encoding="utf8") as f:
    csvreader = csv.reader(f)
    fields = next(csvreader)
    for row in csvreader:
        rows.append(row)
```

2. Then, we print the column names as follows:

```
print(column for column in columns)
```

3. Then, we print the first ten columns:

```
print('First 10 rows:')
for row in rows[:5]:
    for col in row:
        print(col)
    print('\n')
```

This is what the output looks like:

```
First 5 rows:
Juice WRLD
All Girls Are The Same
4VXIryQMWpIdGgYR4TrjT1
83


Schoolgirl Byebye
Year,2015
0UsmyJDsst2xhX1ZiFF3JW
25


Juice WRLD
Lucid Dreams
285pBltuF7vW8TeWk8hdRR
84


Fleetwood Mac
Rhiannon (Will You Ever Win) - 2018 Remaster
4fbwTO3DJ2qryMddov9RbB
49


Joji
SLOW DANCING IN THE DARK
0rKtyWc8bvkriBthvHKY8d
83
```

Figure 4.1 – First five rows of the spotify_data.csv file

How it works...

Have a look at the following code:

```
import csv

filename = "spotify_data.csv"
columns = []
rows = []
```

In the first step of the *How to do it...* section, we imported the built-in library and specified the name of our file. Since it was at the same directory level as our Python script, there was no need to include the full path. Then we declared two lists: one to store our column names (or the first line of our CSV) and the other to store our rows.

Then we proceeded with the `with open` statement. Behind the scenes, the `with` statement creates a context manager that simplifies opening and closing file handlers.

`(filename, 'r')` indicates we want to use `filename` and only read it (`'r'`). After that, we read the file and store the first line in our `columns` list using the `next()` method, which returns the following item from the iterator. For the rest of the records (or rows), we used the `for` iteration to store them in a list.

Since both the declared variables are lists, we can easily read them using the `for` iterator.

There's more...

For this recipe, we used the built-in CSV library of Python and created a simple structure for handling the columns and rows of our CSV file; however, there is a more straightforward way to do it using pandas.

pandas is a Python library that was built to analyze and manipulate data by converting it into structures called **DataFrames**. Don't worry if this is a new concept for you; we will cover it in the following recipes and chapters.

Let's see an example of using pandas to read a CSV file:

1. We need to install pandas. To do this, use the following command:

    ```
    $ pip install pandas
    ```

> **Note**
>
> Remember to use the `pip` command that is associated with your Python version. For some readers, it might be best to use `pip3`. You can verify the version of `pip` and the associated Python version using the `pip -version` command in the CLI.

2. Then, we read the CSV file:

```
import pandas as pd

spotify_df = pd.read_csv('spotify_data.csv')
spotify_df.head()
```

You should have the following output:

	artist_name	track_name	track_id	popularity
0	Juice WRLD	All Girls Are The Same	4VXlryQMWpldGgYR4TrjT1	83
1	Schoolgirl Byebye	Year,2015	0UsmyJDsst2xhX1ZiFF3JW	25
2	Juice WRLD	Lucid Dreams	285pBltuF7vW8TeWk8hdRR	84
3	Fleetwood Mac	Rhiannon (Will You Ever Win) - 2018 Remaster	4fbwTO3DJ2qryMddov9RbB	49
4	Joji	SLOW DANCING IN THE DARK	0rKtyWc8bvkriBthvHKY8d	83

Figure 4.2 – spotify_df DataFrame first five lines

> **Note**
>
> Due to its specific rendering capabilities, this *friendly* visualization can only be seen when using Jupyter Notebook. The output is entirely different if you execute the .head() method on the command line or in your code.

See also

There are other ways to install pandas, and you can explore one here: https://pandas.pydata.org/docs/getting_started/install.html.

Reading a JSON file

JavaScript Object Notation (JSON) is a semi-structured data format. Some articles also define JSON as an unstructured data format, but the truth is this format can be used for multiple purposes.

JSON structure uses nested objects and arrays and, due to its flexibility, many applications and APIs use it to export or share data. That is why describing this file format in this chapter is essential.

This recipe will explore how to read a JSON file using a built-in Python library and explain how the process works.

> **Note**
> JSON is an alternative to XML files, which are very verbose and require more coding to manipulate their data.

Getting ready

This recipe is going to use the GitHub Events JSON data, which can be found in the GitHub repository of this book at `https://github.com/jdorfman/awesome-json-datasets` with other free JSON data.

To retrieve the data, click on **GitHub API | Events**, copy the content from the page, and save it as a `.json` file.

How to do it...

Follow these steps to complete this recipe:

1. Let's start by reading the file:

    ```
    import json
    filename_json = 'github_events.json'

    with open (filename_json, 'r') as f:
        github_events = json.loads(f.read())
    ```

2. Now, let's get the data by making the lines interact with our JSON file:

    ```
    id_list = [item['id'] for item in github_events]
    print(id_list)
    ```

    The output looks as follows:

    ```
    ['25208138097',
     '25208138110',
     (...)
     '25208138008',
     '25208137998']
    ```

How it works...

Have a look at the following code:

```
import json
filename_json = 'github_events.json'
```

As for the CSV file, Python also has a built-in library for JSON files, and we start our script by importing it. Then, we define a variable to refer to our filename.

Have a look at the following code:

```
with open (filename_json, 'r') as f:
    github_events = json.loads(f.read())
```

Using the `open()` function, we open the JSON file. The `json.loads()` statement can't open JSON files. To do that, we used `f.read()`, which will return the file's content as it is, which is then passed as an argument to the first statement.

However, there is a trick here. Unlike in the CSV file, we don't have one single line with the names of the columns. Instead, each data record has its own key representing the data. Since a JSON file is very similar to a Python dictionary, we need to iterate over each record to get all the `id` values in the file.

To simplify the process a bit, we have created the following one-line `for` loop inside a list:

```
id_list = [item['id'] for item in github_events]
```

There's more...

Even though using Python's built-in JSON library seemed very simple, manipulating the data better or filtering by one line, in this case, can create unnecessary complexity. We can use **pandas** once more to simplify the process.

Let's read JSON with pandas. Check out the following code:

```
import pandas as pd

github_events = pd.read_json('github_events.json')
github_events.head(3)
```

The output looks as follows:

id	type	actor	repo	payload	public
25208138097	WatchEvent	{'id': 2286923, 'login': 'hadfield', 'display_...	{'id': 425209123, 'name': 'wenet-e2e/wekws', '...	{'action': 'started'}	True
25208138110	IssueCommentEvent	{'id': 7432848, 'login': 'esclear', 'display_l...	{'id': 88370518, 'name': 'grocy/grocy', 'url'...	{'action': 'created', 'issue': {'url': 'https:...	True
25208138076	PushEvent	{'id': 109922331, 'login': 'ortursucceeh', 'di...	{'id': 564416624, 'name': 'ortursucceeh/Projec...	{'push_id': 11648432335, 'size': 1, 'distinct_...	True

Figure 4.3 – github_events DataFrame first five lines

Then, let's get the id list:

```
github_events['id']
```

The output looks as follows:

```
0        25208138097
1        25208138110
2        25208138076
(...)
27       25208138012
28       25208138008
29       25208137998
Name: id, dtype: int64
```

Like what we did before to read the CSV file, reading a JSON file with pandas was very simple and saved a lot of time and code to get the same information we needed. While in the first example, we needed to iterate over the JSON object list, pandas natively understands it as a DataFrame. Since every column behaves as a one-dimensional array, we can quickly get the values by passing the name of the column (or key) to the DataFrame name.

As with any other library, pandas has limitations, such as reading multiple files simultaneously, parallelism, or reading large datasets. Having this in mind can prevent problems and help us use this library optimally.

Why DataFrames?

DataFrames are bi-dimensional and size-mutable tabular structures and visually resemble a structured table. Due to their versatility, they are widely used in libraries such as pandas (as we saw previously).

PySpark is no different. **Spark** uses DataFrames as distributed data collections and can *parallelize* its tasks to process them through other cores or nodes. We will cover this in more depth in *Chapter 7*.

See also

To find out more about the files supported by the pandas library, follow this link: `https://pandas.pydata.org/pandas-docs/stable/user_guide/io.html`.

Creating a SparkSession for PySpark

Previously introduced in *Chapter 1*, **PySpark** is a Spark library that was designed to work with Python. PySpark uses a Python API to write **Spark** functionalities such as data manipulation, processing (batch or real-time), and machine learning.

However, before ingesting or processing data using PySpark, we must initialize a SparkSession. This recipe will teach us how to create a SparkSession using PySpark and explain its importance.

Getting ready

We first need to ensure we have the correct PySpark version. We installed PySpark in *Chapter 1*; however, checking if we are using the correct version is always good. Run the following command:

```
$ pyspark -version
```

You should see the following output:

```
Welcome to
      ____              __
     / __/__  ___ _____/ /__
    _\ \/ _ \/ _ `/ __/  '_/
   /___/ .__/\_,_/_/ /_/\_\   version 3.1.2
      /_/

Using Scala version 2.12.10, OpenJDK 64-Bit Server VM, 1.8.0_342
Branch HEAD
Compiled by user centos on 2021-05-24T04:27:48Z
Revision de351e30a90dd988b133b3d00fa6218bfcaba8b8
Url https://github.com/apache/spark
Type --help for more information.
```

Next, we choose a code editor that can be any code editor that you want. I will use Jupyter due to the interactive interface.

How to do it...

Let's see how to create a SparkSession:

1. We first create `SparkSession` as follows:

    ```
    from pyspark.sql import SparkSession

    spark = SparkSession.builder \
            .master("local[1]") \
            .appName("DataIngestion") \
            .config("spark.executor.memory", '1g') \
            .config("spark.executor.cores", '3') \
            .config("spark.cores.max", '3') \
            .enableHiveSupport() \
            .getOrCreate()
    ```

    And these are the warnings received:

    ```
    22/11/14 11:09:55 WARN Utils: Your hostname, DESKTOP-DVUDB98
    resolves to a loopback address: 127.0.1.1; using 172.27.100.10
    instead (on interface eth0)
    22/11/14 11:09:55 WARN Utils: Set SPARK_LOCAL_IP if you need to
    bind to another address
    22/11/14 11:09:56 WARN NativeCodeLoader: Unable to load native-
    hadoop library for your platform... using builtin-java classes
    where applicable
    Using Spark's default log4j profile: org/apache/spark/log4j-
    defaults.properties
    Setting default log level to "WARN".
    To adjust logging level use sc.setLogLevel(newLevel). For
    SparkR, use setLogLevel(newLevel).
    ```

    We are not going to have an output when executing the code. Also, don't worry about the WARN messages; they will not affect our work.

2. Then we get the Spark UI. To do this, in your Jupyter cell, type and execute the name of your instantiated as follows:

    ```
    spark
    ```

You should see the following output:

SparkSession - hive
SparkContext

<u>Spark UI</u>
Version
v3.1.2
Master
local[1]
AppName
DataIngestion

Figure 4.4 – Output of execution of instance

3. Next, we access SparkUI in the browser. To do this, we click on the **Spark UI** hyperlink. It will open a new tab in your browser, showing a graph with **Executors** and **Jobs**, and other helpful information:

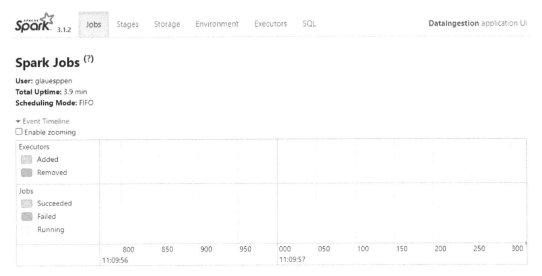

Figure 4.5 – Spark UI home page with an Event Timeline graph

Since we still haven't executed any processes, the graph is empty. Don't worry; we will see it in action in the following recipes.

How it works...

As we saw at the beginning of this chapter, a is a fundamental part of starting our Spark jobs. It sets all the required configurations for **Spark's YARN** (short for **Yet Another Resource Manager**) to allocate the memory, cores, and paths to write temporary and final outputs, among other things.

With this in mind, let's visit each step of our code:

```
spark = .builder \
```

To execute operations such as creating DataFrames, we need to use an instance of . The `spark` variable will be used to access DataFrames and other procedures. Now, take a look at this:

```
    .master("local[1]") \
    .appName("DataIngestion") \
```

The `.master()` method indicates which type of distributed processing we have. Since this is our local machine used only for educational purposes, we defined it as `"local[1]"`, where the integer value needs to be greater than 0, since it represents the number of partitions. The `.appName()` method defines the name of our application session.

After declaring the name of our application, we can set the `.config()` methods:

```
    .config("spark.executor.memory", '1g') \
    .config("spark.executor.cores", '3') \
    .config("spark.cores.max", '3') \
    .enableHiveSupport() \
```

Here, we defined two types of configuration: memory allocation and core allocation. `spark.executor.memory` tells YARN how much memory each executor can allocate to process data; g represents the size unit of it.

`spark.executor.cores` defines the number of executors used by YARN. By default, 1 is used. Next, `spark.cores.max` sets how many cores YARN can scale.

Last, `.enableHiveSupport()` enables the support of Hive queries.

Finally, let's look at this:

```
    .getOrCreate()
```

`.getOrCreate()` is simple as its name. If there is a session with this name, it will retrieve its configurations of it. Otherwise, it will create a new one.

There's more...

Looking at the Spark documentation page, you will see that most configurations are not required to start our jobs or ingestions. However, it is essential to remember that we are handling a scalable framework that was created to allocate resources in a single machine or in clusters to process vast amounts of data. With no limit set to a SparkSession, YARN will allocate every resource it needs to process or ingest data, and this can result in server downtime or even freeze your entire local machine.

In a real-world scenario, a Kubernetes cluster is typically used to ingest or process shared data with other applications or users who do the same thing as you or your team. Physical memory and computational resources tend to be limited, so it is always an excellent practice to set the configurations, even in small projects that use serverless cloud solutions.

Getting all configurations

It is also possible to retrieve the current configurations used for this application session using the following code:

```
spark.sparkContext.getConf().getAll()
```

This is the result we get:

```
[('spark.driver.host', '172.27.107.119'),
 ('spark.executor.id', 'driver'),
 ('spark.sql.warehouse.dir', 'file:/home/glauesppen/spark-warehouse/'),
 ('spark.app.startTime', '1668542489982'),
 ('spark.cores.max', '3'),
 ('spark.app.id', 'local-1668542490406'),
 ('spark.sql.catalogImplementation', 'hive'),
 ('spark.rdd.compress', 'True'),
 ('spark.app.name', 'DataIngestion'),
 ('spark.serializer.objectStreamReset', '100'),
 ('spark.executor.memory', '8g'),
 ('spark.submit.pyFiles', ''),
 ('spark.executor.cores', '3'),
 ('spark.submit.deployMode', 'client'),
 ('spark.driver.port', '46873'),
 ('spark.ui.showConsoleProgress', 'true'),
 ('spark.master', 'local[1]')]
```

Figure 4.6 – configurations set for the recipe

See also

- See all configurations: https://spark.apache.org/docs/latest/configuration.html

- In early versions of Spark, SparkContext was the starting point for working with data and was replaced by in the recent versions. You can read more about it here: https://sparkbyexamples.com/spark/-vs-sparkcontext/.

Using PySpark to read CSV files

As expected, PySpark provides native support for reading and writing CSV files. It also allows data engineers to pass diverse kinds of setups in case the CSV has a different type of delimiter, special encoding, and so on.

In this recipe, we are going to cover how to read CSV files using PySpark using the most common configurations, and we will explain why they are needed.

Getting ready

You can download the CSV dataset for this recipe from Kaggle: `https://www.kaggle.com/datasets/jfreyberg/spotify-chart-data`. We are going to use the same Spotify dataset as in *Chapter 2*.

As in the *Creating a SparkSession for PySpark* recipe, make sure PySpark is installed and running with the latest stable version. Also, using Jupyter Notebook is optional.

How to do it...

Let's get started:

1. We first import and create a SparkSession :

```
from pyspark.sql import
spark = .builder \
        .master("local[1]") \
        .appName("DataIngestion_CSV") \
        .config("spark.executor.memory", '3g') \
        .config("spark.executor.cores", '1') \
        .config("spark.cores.max", '1') \
        .getOrCreate()
```

2. Then, we read the CSV file:

```
df = spark.read.option('header',True).csv('spotify_data.csv')
```

3. Then, we show the data

```
df.show()
```

This is the result we get:

```
+----------------+--------------------+--------------------+----------+
|     artist_name|          track_name|            track_id|popularity|
+----------------+--------------------+--------------------+----------+
|      Juice WRLD|All Girls Are The...|4VXIryQMWpIdGgYR4...|        83|
|Schoolgirl Byebye|          Year,2015|0UsmyJDsst2xhX1Zi...|        25|
|      Juice WRLD|        Lucid Dreams|285pBltuF7vW8TeWk...|        84|
|   Fleetwood Mac|Rhiannon (Will Yo...|4fbwTO3DJ2qryMddo...|        49|
|            Joji|SLOW DANCING IN T...|0rKtyWc8bvkriBthv...|        83|
|        Revenant|           Year 2018|7AF1WfLvu8ESnVrtH...|         0|
|   Morgan Wallen|      Whiskey Glasses|6foY66mWZN0pSRjZ4...|        78|
|   Frank Sinatra|It Was A Very Goo...|1vLPTWPfJSIrqOhNU...|        43|
|        Lil Baby|Drip Too Hard (Li...|78QR3Wp35dqAhFEc2...|        80|
|   Fleetwood Mac|Gypsy - 2018 Rema...|5nTnApD6zNvuHJe0f...|        46|
|   Billie Eilish|lovely (with Khalid)|0u2P5u61voDfwTYjA...|        87|
|    Anthem Lights|K-LOVE Fan Awards...|5rJw9VsPNdfnV9Ar9...|        31|
|     girl in red|we fell in love i...|1BYZxKSf0aTxp8ZFo...|        83|
|   Fleetwood Mac|Tusk - 2018 Remaster|6U0NPtExZu3tKg8ZD...|        48|
|    XXXTENTACION|                SAD!|3ee8Jmje8o58CHK66...|        84|
|    Anthem Lights|K-Love Fan Awards...|00ohIpPn9LkKpeIqh...|        26|
|      Luke Combs|      Beautiful Crazy|2rxQMGVafnNaRaX1R...|        78|
|   Fleetwood Mac|Dreams - 2018 Rem...|4pbO4YIjnPtGJce4q...|        48|
|        Lil Baby|           Yes Indeed|6vN771E9LK6HP2Dew...|        78|
|   The Kiboomers|The Months of the...|58lQgf1Y5gRJRr6S0...|        28|
+----------------+--------------------+--------------------+----------+
only showing top 20 rows
```

Figure 4.7 – spotify_data.csv DataFrame vision using Spark

How it works...

Simple as it is, we need to understand how reading a file with Spark works to make sure it always executes properly:

```
df = spark.read.option('header',True).csv('spotify_data.csv')
```

We attributed a variable to our code statement. This is a good practice when initializing any file reading because it allows us to control the version of the file and, if a change needs to be made, we attribute it to another variable. The name of the variable is also intentional since we are creating our first DataFrame.

Using the .option() method allows us to tell PySpark which type of configuration we want to pass. In this case, we set header to True, which makes PySpark set the first row of the CSV file as the column names. If we didn't pass the required configuration, the DataFrame would look like this:

```
+----------------+--------------------+--------------------+----------+
|             _c0|                 _c1|                 _c2|       _c3|
+----------------+--------------------+--------------------+----------+
|     artist_name|          track_name|            track_id|popularity|
|      Juice WRLD|All Girls Are The...|4VXIryQMWpIdGgYR4...|        83|
|Schoolgirl Byebye|          Year,2015|0UsmyJDsst2xhX1Zi...|        25|
|      Juice WRLD|        Lucid Dreams|285pBltuF7vW8TeWk...|        84|
|   Fleetwood Mac|Rhiannon (Will Yo...|4fbwTO3DJ2qryMddo...|        49|
|            Joji|SLOW DANCING IN T...|0rKtyWc8bvkriBthv...|        83|
|        Revenant|           Year 2018|7AF1WfLvu8ESnVrtH...|         0|
|   Morgan Wallen|     Whiskey Glasses|6foY66mWZN0pSRjZ4...|        78|
|   Frank Sinatra|It Was A Very Goo...|1vLPTWPfJSIrqOhNU...|        43|
|        Lil Baby|Drip Too Hard (Li...|78QR3Wp35dqAhFEc2...|        80|
|   Fleetwood Mac|Gypsy - 2018 Rema...|5nTnApD6zNvuHJe0f...|        46|
|   Billie Eilish|lovely (with Khalid)|0u2P5u6lvoDfwTYjA...|        87|
|    Anthem Lights|K-LOVE Fan Awards...|5rJw9VsPNdfnV9Ar9...|        31|
|      girl in red|we fell in love i...|1BYZxKSf0aTxp8ZFo...|        83|
|   Fleetwood Mac|Tusk - 2018 Remaster|6U0NPtExZu3tKg8ZD...|        48|
|    XXXTENTACION|                SAD!|3ee8Jmje8o58CHK66...|        84|
|    Anthem Lights|K-Love Fan Awards...|00ohIpPn9LkKpeIqh...|        26|
|      Luke Combs|     Beautiful Crazy|2rxQMGVafnNaRaX1R...|        78|
|   Fleetwood Mac|   Dreams - 2018 Rem...|4pbO4YIjnPtGJce4q...|        48|
|        Lil Baby|          Yes Indeed|6vN771E9LK6HP2Dew...|        78|
+----------------+--------------------+--------------------+----------+
only showing top 20 rows
```

Figure 4.8 – spotify_data.csv DataFrame without column names

There's more...

The content of the files can differ, but some setups are welcome when handling CSV files in PySpark. Here, please note that we are going to change the method to `.options()`:

```
df = spark.read.options(header= 'True',
                        sep=',',
                        inferSchema='True') \
            .csv('spotify_data.csv')
```

`header`, `sep`, and `inferSchema` are the most commonly used setups when reading a CSV file. Although CSV stands for **comma-separated values**, it is not hard to find a system or application that exports this file with another type of separator (such as a pipe or a semicolon), and therefore it is useful to have `sep` (which stands for separator) declared.

Let's see an example of an error when reading a CSV that uses a pipe to separate the strings and passes the wrong separator:

```
df = spark.read.options(header= 'True',
                sep=',',
                inferSchema='True') \
        .csv('spotify_data_pipe.csv')
```

This is how the output looks:

```
+-------------------------------------------+
|artist_name|track_name|track_id|popularity|
+-------------------------------------------+
|                      Juice WRLD|All Gi...|
|                      Schoolgirl Byebye...|
|                      Juice WRLD|Lucid ...|
|                      Fleetwood Mac|Rhi...|
|                      Joji|SLOW DANCING...|
|                      Revenant|Year 201...|
|                      Morgan Wallen|Whi...|
|                      Frank Sinatra|It ...|
|                      Lil Baby|Drip Too...|
|                      Fleetwood Mac|Gyp...|
|                      Billie Eilish|lov...|
|                      Anthem Lights|K-L...|
|                      girl in red|we fe...|
|                      Fleetwood Mac|Tus...|
|                      XXXTENTACION|SAD!...|
|                      Anthem Lights|K-L...|
|                      Luke Combs|Beauti...|
|                      Fleetwood Mac|Dre...|
|                      Lil Baby|Yes Inde...|
|                      The Kiboomers|The...|
+-------------------------------------------+
only showing top 20 rows
```

Figure 4.9 – CSV DataFrame reading without a proper sep definition

As you can see, it creates only one column with all the information. But if we pass `sep=' | '`, it will return correctly:

```
+-----------------+--------------------+--------------------+----------+
|      artist_name|          track_name|            track_id|popularity|
+-----------------+--------------------+--------------------+----------+
|       Juice WRLD|All Girls Are The...|4VXIryQMWpIdGgYR4...|        83|
|Schoolgirl Byebye|                Year|2015|0UsmyJDsst2xhX1Zi...|        25|
|       Juice WRLD|        Lucid Dreams|285pBltuF7vW8TeWk...|        84|
|    Fleetwood Mac|Rhiannon (Will Yo...|4fbwTO3DJ2qryMddo...|        49|
|             Joji|SLOW DANCING IN T...|0rKtyWc8bvkriBthv...|        83|
|         Revenant|           Year 2018|7AFlWfLvu8ESnVrtH...|         0|
|    Morgan Wallen|      Whiskey Glasses|6foY66mWZN0pSRjZ4...|        78|
|    Frank Sinatra|It Was A Very Goo...|1vLPTWPfJSIrqOhNU...|        43|
|         Lil Baby|Drip Too Hard (Li...|78QR3Wp35dqAhFEc2...|        80|
|    Fleetwood Mac|Gypsy - 2018 Rema...|5nTnApD6zNvuHJe0f...|        46|
|    Billie Eilish|lovely (with Khalid)|0u2P5u61voDfwTYjA...|        87|
|     Anthem Lights|K-LOVE Fan Awards...|5rJw9VsPNdfnV9Ar9...|        31|
|      girl in red|we fell in love i...|1BYZxKSf0aTxp8ZFo...|        83|
|    Fleetwood Mac|Tusk - 2018 Remaster|6U0NPtExZu3tKg8ZD...|        48|
|      XXXTENTACION|                SAD!|3ee8Jmje8o58CHK66...|        84|
|     Anthem Lights|K-Love Fan Awards...|00ohIpPn9LkKpeIqh...|        26|
|       Luke Combs|     Beautiful Crazy|2rxQMGVafnNaRaX1R...|        78|
|    Fleetwood Mac|Dreams - 2018 Rem...|4pbO4YIjnPtGJce4q...|        48|
|         Lil Baby|          Yes Indeed|6vN771E9LK6HP2Dew...|        78|
|    The Kiboomers|The Months of the...|581Qgf1Y5gRJRr6S0...|        28|
+-----------------+--------------------+--------------------+----------+
only showing top 20 rows
```

Figure 4.10 – CSV DataFrame with the sep definition set to pipe

Other common .options() configurations

There are other complex situations that, if not corrected during ingestion, can result in some problems with the other ETL steps. Here, I am using the listing.csv dataset, which can be found here at http://data.insideairbnb.com/the-netherlands/north-holland/amsterdam/2022-03-08/visualisations/listings.csv:

1. We first read our common configurations:

```
df_broken = spark.read.options(header= 'True', sep=',',
                        inferSchema='True') \
              .csv('listings.csv')
df_broken.show()
```

```
+-----+--------------------+--------+----------+------------------+--------------------+--------+---------+------------------+
-----+--------------+-----------------+-----------+-----------------+---------------------------+----------------+---------------+
-------+--------------------+
|   id|                name| host_id| host_name|neighbourhood_group|        neighbourhood| latitude|longitude|         room_type|p
rice|minimum_nights|number_of_reviews|last_review|reviews_per_month|calculated_host_listings_count|availability_365|number_of_r
eviews_ltm|             license|
+-----+--------------------+--------+----------+------------------+--------------------+--------+---------+------------------+
-----+--------------+-----------------+-----------+-----------------+---------------------------+----------------+---------------+
-------+--------------------+
| 2818|Quiet Garden View...|    3159|    Daniel|              null|Oostelijk Havenge...| 52.36435|  4.94358|      Private room|
49|             3|             285| 2021-11-21|             1.81|                       1|             62|
7|0363 5F3A 5684 67...|
| 20168|Studio with priva...|   59484| Alexander|              null|         Centrum-Oost| 52.36407|  4.89393|      Private room|
106|             1|             339| 2020-04-09|              2.3|                       2|              0|
0|0363 CBB3 2C10 0C...|
| 27886|Romantic, stylish...|   97647|      Flip|              null|         Centrum-West| 52.38761|  4.89188|      Private room|
134|             2|             228| 2022-02-20|             1.84|                       1|             189|
9|0363 974D 4986 74...|
| 28871|Comfortable doubl...|  124245|     Edwin|              null|         Centrum-West| 52.36775|  4.89092|      Private room|
75|             2|             379| 2022-03-05|              2.7|                       2|             146|
43|0363 607B EA74 0B...|
| 29051|Comfortable singl...|  124245|     Edwin|              null|         Centrum-Oost| 52.36584|  4.89111|      Private room|
55|             2|             532| 2022-03-04|             3.98|                       2|             170|
53|0363 607B EA74 0B...|
| 44391|Quiet 2-bedroom A...|  194779|       Jan|              null|         Centrum-Oost| 52.37168|  4.91471|Entire home/apt|
240|             3|              41| 2021-08-28|             0.29|                       1|              0|
```

Figure 4.11 – listing.csv DataFrame

At first glance, all is normal. However, what happens if I try to execute a simple group by using `room_type`?

```
group = df_1.groupBy("room_type").count()
group.show()
```

This is the output we get:

```
+---------------+-----+
|      room_type|count|
+---------------+-----+
|    Shared room|   24|
|            187|    1|
|           null|    7|
|            126|    1|
|     Hotel room|   87|
|            198|    1|
|Entire home/apt| 3823|
|            156|    1|
|            450|    1|
|   Private room| 1791|
|            129|    1|
|             80|    1|
+---------------+-----+
```

Figure 4.12 – Grouping df_broken by room_type

Ignoring `group` by for now, this happens because the file has plenty of escaped quotes and line breaks.

2. Now, let's set the correct `.options()`:

```
df_1 = spark.read.options(header=True, sep=',',
                          multiLine=True, escape='"') \
                 .csv('listings.csv')
group = df_1.groupBy("room_type").count()
group.show()
```

This is the result:

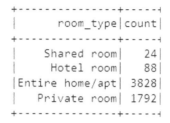

```
+---------------+-----+
|      room_type|count|
+---------------+-----+
|    Shared room|   24|
|     Hotel room|   88|
|Entire home/apt| 3828|
|   Private room| 1792|
+---------------+-----+
```

Figure 4.13 – Grouping by room_type using the right options() settings

See also

You can see more PySpark `.options()` in the official documentation: `https://spark.apache.org/docs/latest/sql-data-sources-csv.html`.

Using PySpark to read JSON files

In the *Reading a JSON file* recipe, we saw that JSON files are widely used to transport and share data between applications, and we saw how to read a JSON file using simple Python code.

However, with the increase in data size and sharing, using only Python to process a high volume of data can lead to performance or resilience issues. That's why, for this type of scenario, it is highly recommended to use PySpark to read and process JSON files. As you might expect, PySpark comes with a straightforward reading solution.

In this recipe, we will cover how to read a JSON file with PySpark, the common associated issues, and how to solve them.

Getting ready

As in the previous recipe, *Reading a JSON file*, we are going to use the GitHub Events JSON file. Also, the use of Jupyter Notebook is optional.

How to do it...

Here are the steps for this recipe:

1. We first create the SparkSession:

```
spark = .builder \
        .master("local[1]") \
        .appName("DataIngestion_JSON") \
        .config("spark.executor.memory", '3g') \
        .config("spark.executor.cores", '1') \
        .config("spark.cores.max", '1') \
        .getOrCreate()
```

2. Then, we read the JSON file:

```
df_json = spark.read.option("multiline", "true") \
                    .json('github_events.json')
```

3. Then, we show the data:

```
df_json.show()
```

This is how the output appears:

```
+--------------------+--------------------+------------+--------------------+--------------------+------+--------------------+--
----------------+
|             actor|          created_at|          id|                 org|             payload|public|                repo|
type|
+--------------------+--------------------+------------+--------------------+--------------------+------+--------------------+--
----------------+
|{https://avatars....|2022-11-13T22:52:04Z|25208138097|{https://avatars....|{started, null, n...| true|{425209123, wenet...|
WatchEvent|
|{https://avatars....|2022-11-13T22:52:04Z|25208138110|{https://avatars....|{created, null, {...| true|{88370518, grocy/...|Is
sueCommentEvent|
|{https://avatars....|2022-11-13T22:52:04Z|25208138076|                null|{null, d89e1fda20...| true|{564416624, ortur...|
PushEvent|
|{https://avatars....|2022-11-13T22:52:04Z|25208138082|                null|{null, null, null...| true|{565604064, mikey...|
CreateEvent|
|{https://avatars....|2022-11-13T22:52:04Z|25208138077|{https://avatars....|{null, 026c8289f4...| true|{346488231, Sud-A...|
PushEvent|
|{https://avatars....|2022-11-13T22:52:04Z|25208138087|                null|{opened, null, nu...| true|{556443902, Strac...| P
ullRequestEvent|
|{https://avatars....|2022-11-13T22:52:04Z|25208138078|                null|{null, 72bad7db76...| true|{364716587, induc...|
PushEvent|
|{https://avatars....|2022-11-13T22:52:04Z|25208138069|                null|{null, b26e119401...| true|{478275335, Thoma...|
PushEvent|
|{https://avatars....|2022-11-13T22:52:04Z|25208138072|{https://avatars....|{closed, null, nu...| true|{32278338, simula...| P
ullRequestEvent|
|{https://avatars....|2022-11-13T22:52:04Z|25208138080|{https://avatars....|{closed, null, nu...| true|{114165574, Maaya...|
IssuesEvent|
|{https://avatars....|2022-11-13T22:52:04Z|25208138070|                null|{opened, null, nu...| true|{280576947, quock...| P
ullRequestEvent|
|{https://avatars....|2022-11-13T22:52:04Z|25208138067|                null|{null, 3f3099cfea...| true|{551586330, Mario...|
PushEvent|
```

Figure 4.14 – df_json DataFrame

How it works...

Similar to CSV files, reading a JSON file using PySpark is very simple, requiring only one line of code. Like pandas, it ignores the brackets in the file and creates a table-structured DataFrame, even though we are handling a semi-structured data file. However, the great magic is in .option("multiline", "true").

If you remember our JSON structure for this file, it is something like this:

```
[
  {
    "id": "25208138097",
    "type": "WatchEvent",
    "actor": {...},
    "repo": {...},
    "payload": {
      "action": "started"
    },
    "public": true,
    "created_at": "2022-11-13T22:52:04Z",
    "org": {...}
  },
  {
    "id": "25208138110",
    "type": "IssueCommentEvent",
    "actor": {...},
    "repo": {...},
  },...
```

It is a multi-lined JSON since it has objects inside objects. The .option() setup passed when reading the file guarantees PySpark will read it as it should, and if we don't pass this argument, an error like this will appear:

```
AnalysisException: Since Spark 2.3, the queries from raw JSON/CSV files are dis
allowed when the
referenced columns only include the internal corrupt record column
(named _corrupt_record by default). For example:
spark.read.schema(schema).json(file).filter($"_corrupt_record".isNotNull).count
()
and spark.read.schema(schema).json(file).select("_corrupt_record").show().
Instead, you can cache or save the parsed results and then send the same query.
For example, val df = spark.read.schema(schema).json(file).cache() and then
df.filter($"_corrupt_record".isNotNull).count().
```

Figure 4.15 – This error is shown when .options() is not well defined for a multi-lined JSON file

PySpark understands it as a corrupted file.

There's more...

A very widely used configuration for reading JSON files is `dropFieldIfAllNull`. When set to `true`, if there is an empty array, it will drop it from the schema.

Unstructured and semi-structured data are valuable due to their elasticity. Sometimes, applications can change their output and some fields become deprecated. To avoid changing the ingest script (especially if these changes can be frequent), `dropFieldIfAllNull` removes them from the DataFrame.

See also

- To find out more about PySpark `.options()`, refer to the official documentation: `https://spark.apache.org/docs/latest/sql-data-sources-json.html`

Further reading

- `https://docs.fileformat.com/spreadsheet/csv/`
- `https://codefather.tech/blog/python-with-open/`
- `https://www.programiz.com/python-programming/methods/built-in/next`
- `https://spark.apache.org/docs/latest/sql-data-sources-csv.html`
- `https://sparkbyexamples.com/pyspark/pyspark-read-json-file-into-dataframe/`

5

Ingesting Data from Structured and Unstructured Databases

Nowadays, we can store and retrieve data from multiple sources, and the optimal storage method depends on the type of information being processed. For example, most APIs make data available in an unstructured format as this allows the sharing of data of multiple formats (for example, audio, video, and image) and has low storage costs via the use of data lakes. However, if we want to make quantitative data available for use with several tools to support analysis, then the most reliable option might be structured data.

Ultimately, whether you are a data analyst, scientist, or engineer, it is essential to understand how to manage both structured and unstructured data.

In this chapter, we will cover the following recipes:

- Configuring a JDBC connection
- Ingesting data from a JDBC database using SQL
- Connecting to a NoSQL database (MongoDB)
- Creating our NoSQL table in MongoDB
- Ingesting data from MongoDB using PySpark

Technical requirements

You can find the code from this chapter in the GitHub repository at `https://github.com/PacktPublishing/Data-Ingestion-with-Python-Cookbook`.

Using the **Jupyter Notebook** is not mandatory but allows us to explore the code interactively. Since we will execute both Python and PySpark code, Jupyter can help us to understand the scripts better. Once you have Jupyter installed, you can execute it using the following line:

```
$ jupyter notebook
```

It is recommended to create a separate folder to store the Python files or notebooks we will cover in this chapter; however, feel free to organize it in the most appropriate way for you.

Configuring a JDBC connection

Working with different systems brings the challenge of finding an efficient way to connect the systems. An adaptor, or a driver, is the solution to this communication problem, creating a bridge to translate information from one system to another.

JDBC, or **Java Database Connectivity**, is used to facilitate communication between Java-based systems and databases. This recipe covers configuring JDBC in SparkSession to connect to a PostgreSQL database, using best practices as always.

Getting ready

Before configuring SparkSession, we need to download the `.jars` file (Java Archive). You can do this at `https://jdbc.postgresql.org/` on the PostgreSQL official site.

Select **Download**, and you will be redirected to another page:

Figure 5.1 – PostgreSQL JDBC home page

Then, select the **Java 8 Download** button.

Keep this `.jar` file somewhere safe, as you will need it later. I suggest keeping it inside the folder where your code is.

For the PostgreSQL database, you can use a Docker image or the instance we created on Google Cloud in *Chapter 4*. If you opt for the Docker image, ensure it is up and running.

The final preparatory step for this recipe is to import a dataset to be used. We will use the `word_population.csv` file (which you can find in the GitHub repository of this book, at `https://github.com/PacktPublishing/Data-Ingestion-with-Python-Cookbook/tree/main/Chapter_5/datasets`). Import it using DBeaver or any other SQL IDE of your choice. We will use this dataset with SQL in the *Ingesting data from a JDBC database using SQL* recipe later in this chapter.

To import data into DBeaver, create a table with the name of your choice under the Postgres database. I chose to give my table the exact name of the CSV file. You don't need to insert any columns for now.

Then, right-click on the table and select **Import Data**, as shown in the following screenshot:

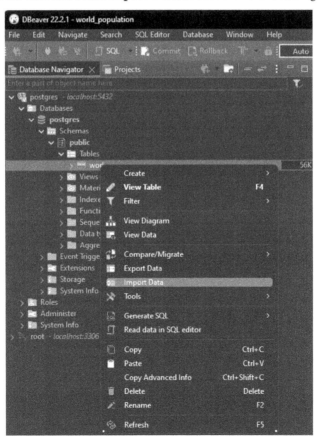

Figure 5.2 – Importing data on a table using DBeaver

A new window will open, showing the options to use a CSV file or a database table. Select **CSV** and then **Next**, as follows:

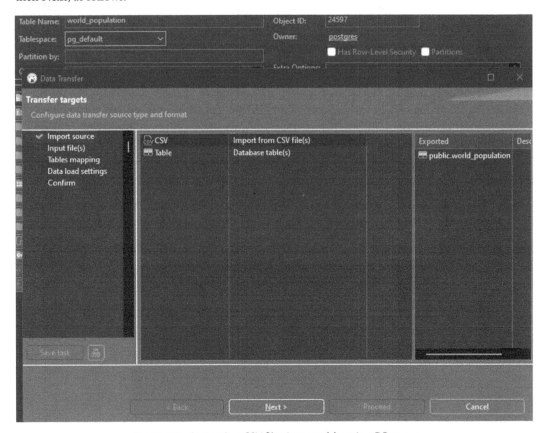

Figure 5.3 – Importing CSV files into a table using DBeaver

A new window will open where you can select the file. Choose the `world_population.csv` file and select the **Next** button, leaving the default settings shown as follows:

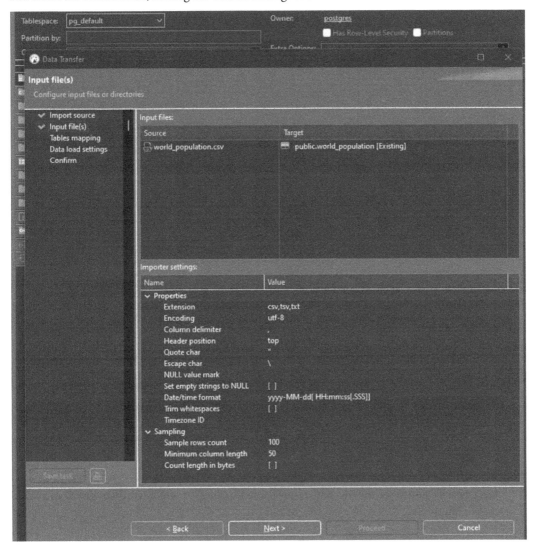

Figure 5.4 – CSV file successfully imported into the world_population table

If all succeeds, you should be able to see the `world_population` table populated with the columns and data:

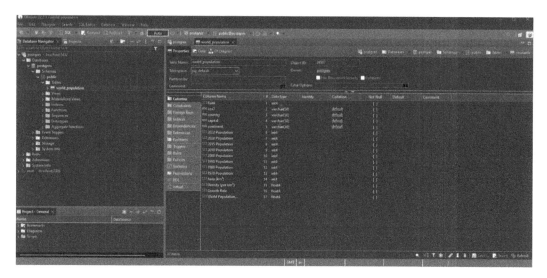

Figure 5.5 – The world_population table populated with data from the CSV

How to do it...

I will use a Jupyter notebook to insert and execute the code to make this exercise more dynamic. Here is how we do it:

1. **Importing the PySpark libraries**: Besides `SparkSession`, we will need an additional class called `SparkConf` to set our new configuration:

   ```
   from pyspark.conf import SparkConf
   from pyspark.sql import SparkSession
   ```

2. **Using SparkConf to set the .jar path**: Using `SparkConf()`, which we instantiated, we can set the path to the `.jar` with `spark.jars`:

   ```
   conf = SparkConf()
   conf.set('spark.jars', /path/to/your/postgresql-42.5.1.jar')
   ```

   You will see a `SparkConf` object created, as shown in the following output:

   ```
   Out[2]:  <pyspark.conf.SparkConf at 0x7f41fc425ac0>
   ```

   Figure 5.6 – SparkConf object

3. **Creating the SparkSession instance**: The next step is to pass the new configuration to `SparkSession` and create it:

```
spark = SparkSession.builder \
        .config(conf=conf) \
        .master("local") \
        .appName("Postgres Connection Test") \
        .getOrCreate()
```

If a warning message appears as in the following screenshot, you can ignore it:

```
22/12/15 21:05:53 WARN Utils: Your hostname, DESKTOP-DVUDB98 resolves to a loopback address: 127.0.1.1; using 172.30.194.211 in
stead (on interface eth0)
22/12/15 21:05:53 WARN Utils: Set SPARK_LOCAL_IP if you need to bind to another address
22/12/15 21:05:55 WARN NativeCodeLoader: Unable to load native-hadoop library for your platform... using builtin-java classes w
here applicable
Using Spark's default log4j profile: org/apache/spark/log4j-defaults.properties
Setting default log level to "WARN".
To adjust logging level use sc.setLogLevel(newLevel). For SparkR, use setLogLevel(newLevel).
```

Figure 5.7 – SparkSession initialization warning messages

4. **Connecting to our database**: Finally, we can connect to the PostgreSQL database by passing the required credentials including host, database name, username, and password as follows:

```
df= spark.read.format("jdbc") \
    .options(url="jdbc:postgresql://localhost:5432/postgres",
            dbtable="world_population",
            user="root",
            password="root",
            driver="org.postgresql.Driver") \
    .load()
```

If the credentials are correct, we should expect no output here.

5. **Getting the schema of the DataFrame**: Using PySpark `.printSchema()`, it is possible now to see the table columns:

```
df.printSchema()
```

Executing the code will show the following output:

```
root
 |-- Rank: integer (nullable = true)
 |-- cca3: string (nullable = true)
 |-- country: string (nullable = true)
 |-- capital: string (nullable = true)
 |-- continent: string (nullable = true)
 |-- 2022 Population: integer (nullable = true)
 |-- 2020 Population: integer (nullable = true)
 |-- 2015 Population: integer (nullable = true)
 |-- 2010 Population: integer (nullable = true)
 |-- 2000 Population: integer (nullable = true)
 |-- 1990 Population: integer (nullable = true)
 |-- 1980 Population: integer (nullable = true)
 |-- 1970 Population: integer (nullable = true)
 |-- Area (km²): integer (nullable = true)
 |-- Density (per km²): float (nullable = true)
 |-- Growth Rate: float (nullable = true)
 |-- World Population Percentage: float (nullable = true)
```

Figure 5.8 – DataFrame of the world_population schema

How it works...

We can observe that PySpark (and Spark) require additional configuration to create a connection with a database. In this recipe, using the PostgreSQL .jars file is essential to make it work.

Let's understand what kind of configuration Spark requires by looking at our code:

```
conf = SparkConf()
conf.set('spark.jars', '/path/to/your/postgresql-42.5.1.jar')
```

We started by instantiating the SparkConf() method, responsible for defining configurations used in SparkSession. After instantiating the class, we used the set() method to pass a key-value pair parameter: spark.jars. If more than one .jars file was used, the paths could be passed on the value parameter separated by commas. It is also possible to define more than one conf.set() method; they just need to be included one after the other.

It is on the second line of SparkSession where the set of configurations is passed, as you can see in the following code:

```
spark = SparkSession.builder \
        .config(conf=conf) \
        .master("local") \
    (...)
```

Then, with our SparkSession instantiated, we can use it to read our database, as you can see in the following code:

```
df= spark.read.format("jdbc") \
    .options(url="jdbc:postgresql://localhost:5432/postgres",
            dbtable="world_population",
            user="root",
            password="root",
            driver="org.postgresql.Driver") \
    .load()
```

Since we are handling a third-party application, we must set the format for reading the output using the .format() method. The .options() method will carry the authentication values and the driver.

> **Note**
>
> With time you will observe that there are a few diverse ways to declare the .options() key-value pairs. For example, another frequently used format is .options("driver", "org.postgresql.Driver). Both ways are correct depending on the *taste* of the developer.

There's more...

This recipe covered how to use a JDBC driver, and the same logic applies to **Open Database Connectivity (ODBC)**. However, determining the criteria for using JDBC or ODBC requires understanding which data source we are ingesting data from.

The ODBC connection in Spark is usually associated with Spark Thrift Server, a Spark SQL extension from Apache HiveServer2 that allows users to execute SQL queries in **Business Intelligence (BI)** tools such as MS PowerBI or Tableau. See the following diagram for an outline of this relationship:

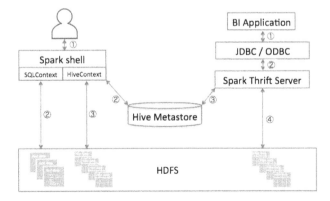

Figure 5.9 – Spark Thrift architecture, provided by Cloudera documentation (https://docs.cloudera.com/ HDPDocuments/HDP3/HDP-3.1.5/developing-spark-applications/content/using_spark_sql.html)

By contrast to JDBC, ODBC is used in real-life projects that are smaller and more specific to certain system integrations. It also requires the use of another Python library called pyodbc. You can read more about it at `https://kontext.tech/article/290/connect-to-sql-server-in-spark-pyspark`.

Debugging connection errors

PySpark errors can be very confusing and lead to misinterpretations. It happens because the errors are often related to a problem on the JVM, and Py4J (a Python interpreter that communicates dynamically with the JVM) consolidates the message with other Python errors that may have occurred.

Some error messages are prevalent and can easily be identified when managing database connections. Let's take a look at an error that occurred when using the following code:

```
df= spark.read.format("jdbc") \
    .options(url="jdbc:postgresql://localhost:5432/postgres",
            dbtable="world_population",
            user="root",
            password="root") \
    .load()
```

Here is the error message that resulted:

```
Py4JJavaError: An error occurred while calling o34.load.
: java.sql.SQLException: No suitable driver
        at java.sql/java.sql.DriverManager.getDriver(DriverManager.java:300)
        at org.apache.spark.sql.execution.datasources.jdbc.JDBCOptions.$anonfun$driverClass$2(JDBCOptions.scala:107)
        at scala.Option.getOrElse(Option.scala:189)
        at org.apache.spark.sql.execution.datasources.jdbc.JDBCOptions.<init>(JDBCOptions.scala:107)
        at org.apache.spark.sql.execution.datasources.jdbc.JDBCOptions.<init>(JDBCOptions.scala:39)
        at org.apache.spark.sql.execution.datasources.jdbc.JdbcRelationProvider.createRelation(JdbcRelationProvider.s
cala:34)
        at org.apache.spark.sql.execution.datasources.DataSource.resolveRelation(DataSource.scala:350)
        at org.apache.spark.sql.DataFrameReader.loadV1Source(DataFrameReader.scala:228)
        at org.apache.spark.sql.DataFrameReader.$anonfun$load$2(DataFrameReader.scala:210)
        at scala.Option.getOrElse(Option.scala:189)
        at org.apache.spark.sql.DataFrameReader.load(DataFrameReader.scala:210)
        at org.apache.spark.sql.DataFrameReader.load(DataFrameReader.scala:171)
        at java.base/jdk.internal.reflect.DirectMethodHandleAccessor.invoke(DirectMethodHandleAccessor.java:104)
        at java.base/java.lang.reflect.Method.invoke(Method.java:578)
        at py4j.reflection.MethodInvoker.invoke(MethodInvoker.java:244)
        at py4j.reflection.ReflectionEngine.invoke(ReflectionEngine.java:357)
        at py4j.Gateway.invoke(Gateway.java:282)
        at py4j.commands.AbstractCommand.invokeMethod(AbstractCommand.java:132)
        at py4j.commands.CallCommand.execute(CallCommand.java:79)
        at py4j.ClientServerConnection.waitForCommands(ClientServerConnection.java:182)
        at py4j.ClientServerConnection.run(ClientServerConnection.java:106)
        at java.base/java.lang.Thread.run(Thread.java:1589)
```

Figure 5.10 – Py4JJavaError message

In the first line, we see `Py4JJavaError` informing us of an error when calling the load function. Continuing to the second line, we can see the message: `java.sql.SQLException: No suitable driver`. It informs us that even though the `.jars` file is configured and set, PySpark doesn't know which drive to use to load data from PostgreSQL. This can be easily fixed by adding the `driver` parameter under `.options()`. Refer to the following code:

```
df= spark.read.format("jdbc") \
    .options(url="jdbc:postgresql://localhost:5432/postgres",
            dbtable="world_population",
            user="root",
            password="root",
            driver="org.postgresql.Driver") \
    .load()
```

See also

Find more about Spark Thrift Server at `https://jaceklaskowski.gitbooks.io/mastering-spark-sql/content/spark-sql-thrift-server.html`.

Ingesting data from a JDBC database using SQL

With the connection tested and SparkSession configured, the next step is to ingest the data from PostgreSQL, filter it, and save it in an analytical format called a Parquet file. Don't worry about how Parquet files work for now; we will cover it in the following chapters.

This recipe aims to use the connection we created with our JDBC database and ingest the data from the `world_population` table.

Getting ready

This recipe will use the same dataset and code as the *Configuring a JDBC connection* recipe to connect to the PostgreSQL database. Ensure your Docker container is running or your PostgreSQL server is up.

This recipe continues from the content presented in *Configuring a JDBC connection*. We will now learn how to ingest the data inside the Postgres database.

How to do it...

Following on from our previous code, let's read the data in our database as follows:

1. **Creating our DataFrame**: Using the previous connection settings, let's read the data from PostgreSQL using the `world_population` table:

    ```
    df= spark.read.format("jdbc") \
        .options(url="jdbc:postgresql://localhost:5432/postgres",
                dbtable="world_population",
                user="root",
                password="root",
                driver="org.postgresql.Driver") \
        .load()
    ```

2. **Creating a TempView**: Using the exact name of our table (for organization purposes), we create a temporary view in the Spark default database from the DataFrame:

    ```
    df.createOrReplaceTempView("world_population")
    ```

    There is no output expected here.

3. **Using SQL to filter data**: With the temporary view, we can use the SQL function from SparkSession, instantiated by the `spark` variable:

    ```
    spark.sql("select * from world_population").show(3)
    ```

    Depending on the size of your monitor, the output may look confusing, as follows:

```
+----+----+-----------+-------+---------+---------------+---------------+---------------+---------------+-----------
---+---------------+---------------+---------------+---------+----------+---------------+-----------+--------------------
---+
|Rank|cca3|    country|capital|continent|2022 Population|2020 Population|2015 Population|2010 Population|2000 Populat
ion|1990 Population|1980 Population|1970 Population|Area (km²)|Density (per km²)|Growth Rate|World Population Percent
age|
+----+----+-----------+-------+---------+---------------+---------------+---------------+---------------+-----------
---+---------------+---------------+---------------+---------+----------+---------------+-----------+--------------------
---+
|  36| AFG|Afghanistan|  Kabul|     Asia|       41128771|       38972230|       33753499|       28189672|       19542
982|       10694796|       12486631|       10752971|    652230|         63.0587|     1.0257|
0.52|
| 138| ALB|    Albania| Tirana|   Europe|        2842321|        2866849|        2882481|        2913399|        3182
021|        3295066|        2941651|        2324731|     28748|         98.8702|     0.9957|
0.04|
|  34| DZA|    Algeria|Algiers|   Africa|       44903225|       43451666|       39543154|       35856344|       30774
621|       25518074|       18739378|       13795915|   2381741|         18.8531|     1.0164|
0.56|
+----+----+-----------+-------+---------+---------------+---------------+---------------+---------------+-----------
---+---------------+---------------+---------------+---------+----------+---------------+-----------+--------------------
---+
only showing top 3 rows
```

Figure 5.11 – world_population view using Spark SQL

4. **Filtering data**: Using a SQL statement, let's filter only the South American countries in our DataFrame:

```
south_america = spark.sql("select * from world_population where
continent = 'South America' ")
```

Since we attribute the results to a variable, there is no output.

5. **Using toPandas()**: To ensure this is the correct information that we want to ingest, let's use the `.toPandas()` function to bring a more user-friendly view:

```
south_america.toPandas()
```

This is how the result appears:

	Rank	cca3	country	capital	continent	2022 Population	2020 Population	2015 Population	2010 Population	2000 Population	1990 Population	1980 Population	1970 Population	Area (km²)	D (pe
0	33	ARG	Argentina	Buenos Aires	South America	45510318	45036032	43257065	41100123	37070774	32637657	28024803	23842803	2780400	16.3
1	80	BOL	Bolivia	Sucre	South America	12224110	11936162	11090085	10223270	8592656	7096194	5736088	4585693	1098581	11.1
2	7	BRA	Brazil	Brasília	South America	215313498	213196304	205188205	196353492	175873720	150706446	122288383	96369875	8515767	25.2
3	65	CHL	Chile	Santiago	South America	19603733	19300315	17870124	17004162	15351799	13342868	11469828	9820481	756102	25.9
4	28	COL	Colombia	Bogota	South America	51874024	50930662	47119728	44816108	39215135	32601393	26176195	20905254	1141748	45.4
5	67	ECU	Ecuador	Quito	South America	18001000	17588595	16195902	14989585	12626507	10449837	8135845	6172215	276841	65.0
6	231	FLK	Falkland Islands	Stanley	South America	3780	3747	3408	3187	3080	2332	2240	2274	12173	0.3
7	184	GUF	French Guiana	Cayenne	South America	304557	290969	257026	228453	164351	113931	66825	46484	83534	3.6
8	164	GUY	Guyana	Georgetown	South America	808726	797202	755031	747932	759051	747116	778176	705261	214969	3.7
9	109	PRY	Paraguay	Asunción	South America	6780744	6618695	6177950	5768613	5123819	4059195	3078912	2408787	406752	16.6
10	44	PER	Peru	Lima	South America	34049588	33304756	30711863	29229572	26654439	22109099	17492406	13562371	1285216	26.4
11	170	SUR	Suriname	Paramaribo	South America	618040	607065	575475	546080	478998	412756	375112	379918	163820	3.7
12	133	URY	Uruguay	Montevideo	South America	3422794	3429086	3402818	3352651	3292224	3117012	2953750	2790265	181034	18.9
13	51	VEN	Venezuela	Caracas	South America	28301696	28490453	30529716	28715022	24427729	19750579	15210443	11355475	916445	30.8

Figure 5.12 – south_america countries with toPandas() visualization

6. **Saving our work**: Now, we can save our filtered data as follows:

```
south_america.write.parquet('south_america_population')
```

Looking at your script's folder, you should see a folder named `south_america_population`. Inside, you should see the following output:

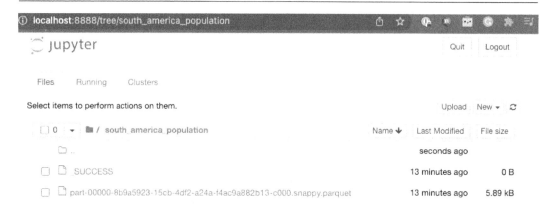

Figure 5.13 – south_america data in the Parquet file

This is our filtered and ingested DataFrame in an analytical format.

How it works...

A significant advantage of working with Spark is the possibility of using SQL statements to filter and query data from a DataFrame. It allows data analytics and BI teams to help the data engineers by handling queries. This helps to build analytical data and insert it into data warehouses.

Nevertheless, there are some considerations we need to take to execute a SQL statement properly. One of them is using `.createOrReplaceTempView()`, as seen in this line of code:

```
df.createOrReplaceTempView("world_population")
```

Behind the scenes, this temporary view will work as a SQL table and organize the data from the DataFrame without needing physical files.

Then we used the instantiated `SparkSession` variable to execute the SQL statements. Note that the name of the table is the same as the temporary view:

```
spark.sql("select * from world_population").show(3)
```

After doing the SQL queries we required, we proceeded to save our files using the `.write()` method, as follows:

```
south_america.write.parquet('south_america_population')
```

The parameter inside the `parquet()` method defines the file's path and name. Several other configurations are available when writing Parquet files, which we will cover later, in *Chapter 7*.

There's more...

Although we used a temporary view to make our SQL statements, it is also possible to use the filtering and aggregation functions from the DataFrame. Let's use the example from this recipe by filtering only the South American countries:

```
df.filter(df['continent'] == 'South America').show(10)
```

You should see the following output:

```
+----+----+----------------+------------+--------------+--------------+--------------+--------------
-+--------------+--------------+--------------+--------------+----------+---------------+--------------
----------------+
|Rank|cca3|         country|     capital|     continent|2022 Population|2020 Population|2015 Population|2010 Populatio
n|2000 Population|1990 Population|1980 Population|1970 Population|Area (km²)|Density (per km²)|Growth Rate|World Popu
lation Percentage|
+----+----+----------------+------------+--------------+--------------+--------------+--------------
-+--------------+--------------+--------------+--------------+----------+---------------+--------------
----------------+
|  33| ARG|       Argentina|Buenos Aires|South America|      45510318|      45036032|      43257065|      4110012
3|       37070774|       32637657|       28024803|      23842803|   2780400|        16.3683|     1.0052|
0.57|
|  80| BOL|         Bolivia|       Sucre|South America|      12224110|      11936162|      11090085|      1022327
0|        8592656|        7096194|        5736088|       4585693|   1098581|        11.1272|      1.012|
0.15|
|   7| BRA|          Brazil|    Brasilia|South America|     215313498|     213196304|     205188205|     19635349
2|      175873720|      150706446|      122288383|      96369875|   8515767|        25.2841|     1.0046|
2.7|
|  65| CHL|           Chile|    Santiago|South America|      19603733|      19300315|      17870124|      1700416
2|       15351799|       13342868|       11469828|       9820481|    756102|        25.9274|     1.0057|
0.25|
|  28| COL|        Colombia|      Bogota|South America|      51874024|      50930662|      47119728|      4481610
8|       39215135|       32601393|       26176195|      20905254|   1141748|        45.4339|     1.0069|
0.65|
|  67| ECU|         Ecuador|       Quito|South America|      18001000|      17588595|      16195902|      1498958
5|       12626507|       10449837|        8135845|       6172215|    276841|        65.0229|     1.0114|
0.23|
| 231| FLK|Falkland Islands|     Stanley|South America|          3780|          3747|          3408|           318
7|           3080|           2332|           2240|          2274|     12173|         0.3105|     1.0043|
0.0|
| 184| GUF|   French Guiana|     Cayenne|South America|        304557|        290969|        257026|         22845
3|         164351|         113931|          66825|         46484|     83534|         3.6459|     1.0239|
0.0|
| 164| GUY|          Guyana|  Georgetown|South America|        808726|        797202|        755031|         74793
2|         759051|         747116|         778176|        705261|    214969|         3.7621|     1.0052|
0.01|
| 109| PRY|        Paraguay|    Asunción|South America|       6780744|       6618695|       6177950|        576861
3|        5123819|        4059195|        3078912|       2408787|    406752|        16.6705|     1.0115|
0.09|
+----+----+----------------+------------+--------------+--------------+--------------+--------------
-+--------------+--------------+--------------+--------------+----------+---------------+--------------
----------------+
only showing top 10 rows
```

Figure 5.14 – South American countries filtered using DataFrame operations

It is essential to understand that not all SQL functions can be used as DataFrame operations. You can see more practical examples of filtering and aggregation functions using DataFrame operations at https://spark.apache.org/docs/2.2.0/sql-programming-guide.html.

See also

TowardsDataScience has a fantastic blog post about SQL functions using PySpark, at https://towardsdatascience.com/pyspark-and-sparksql-basics-6cb4bf967e53.

Connecting to a NoSQL database (MongoDB)

MongoDB is an open source, unstructured, document-oriented database made in C++. It is well known in the data world for its scalability, flexibility, and speed.

As someone who will work with data (or maybe already does), it is essential to know how to explore a MongoDB (or any other unstructured) database. MongoDB has some peculiarities, which we will explore practically here.

In this recipe, you will learn how to create a connection to access MongoDB documents via Studio 3T Free, a MongoDB GUI.

Getting ready

To start our work with this robust database, first, we need to install and create a MongoDB server on our local machine. We already configured a MongoDB Docker container in *Chapter 1*, so let's get it up and running. You can do this using Docker Desktop or via the command line using the following command:

```
my-project/mongo-local$ docker run \
--name mongodb-local \
-p 27017:27017 \
-e MONGO_INITDB_ROOT_USERNAME=<your_username> \
-e MONGO_INITDB_ROOT_PASSWORD=<your_password>\
-d mongo:latest
```

Don't forget to change the variables using the username and password of your choice.

On Docker Desktop, you should see the following:

Figure 5.15 – MongoDB Docker container running

The next step is to download and configure Studio 3T Free, free software the development community uses to connect to MongoDB servers. You can download this software from https://studio3t. com/download-studio3t-free and follow the installer's steps for your given OS.

During the installation, a message may appear like that shown in the following figure. If so, you can leave the fields blank. We don't need password encryption for local or testing purposes.

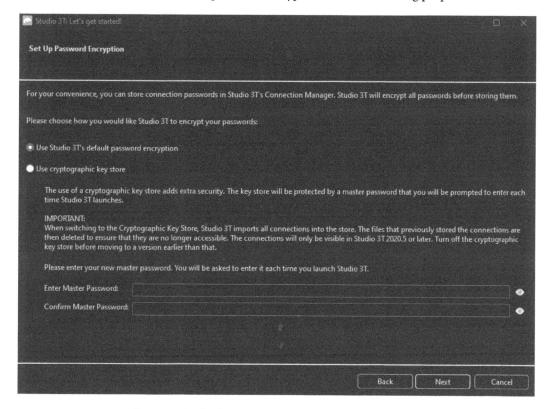

Figure 5.16 – Studio 3T Free password encryption message

When the installation process is finished, you will see the following window:

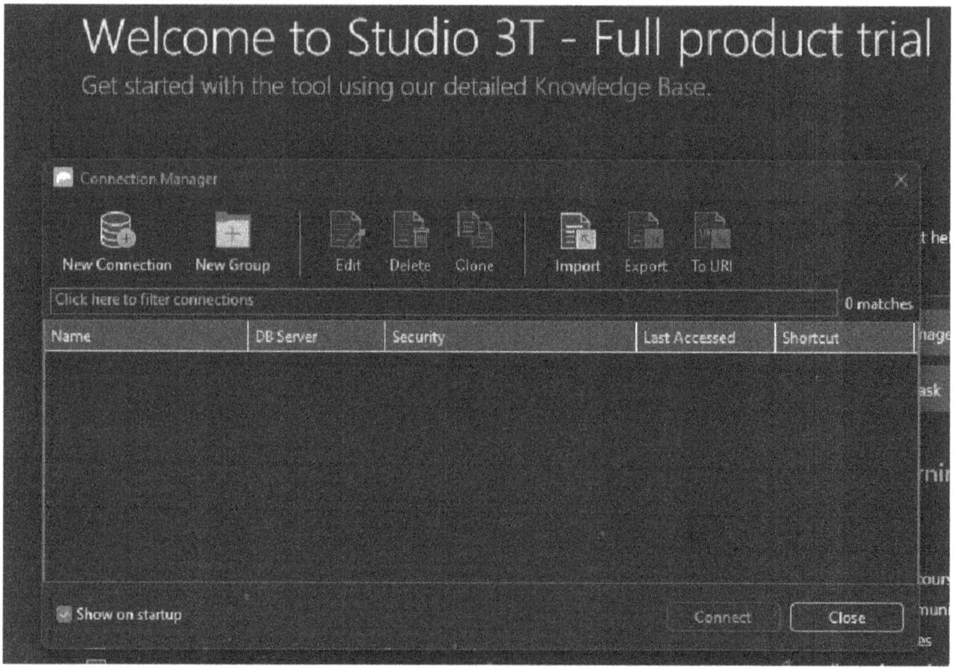

Figure 5.17 – Studio 3T Free connection window

We are now ready to connect our MongoDB instance to the IDE.

How to do it...

Now that we have Studio 3T installed, let's connect to our local MongoDB instance:

1. **Creating the connection**: Right after you open Studio 3T, a window will appear and ask you to insert the connection string or manually configure it. Select the second option and click on **Next**.

You will have something like this:

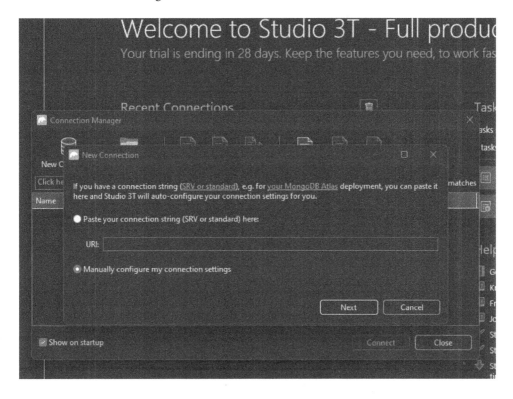

Figure 5.18 – Studio 3T New Connection initial options

2. **Inserting the connection information**: In the new window that appears, give your database connection an appropriate name in the **Connection name** field. Leave the information filled in by default by Studio 3T, where **Connection Type** is set to localhost and **Port** to 27017.

Your screen should look as follows for now:

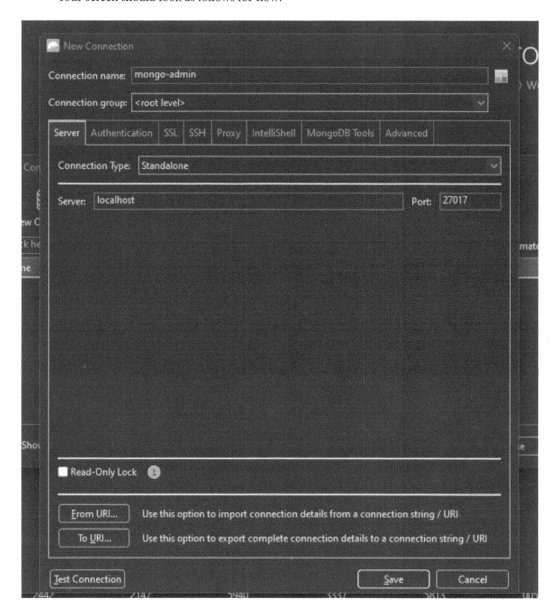

Figure 5.19 – New Connection server information

Now select the **Authentication** tab under the **Connection group** field, and from the **Authentication Mode** drop-down menu, choose **Basic**.

Three fields will appear—**User name**, **Password**, and **Authentication DB**. Fill them in with the credentials you used when creating the Docker image for MongoDB, and enter admin in the **Authentication DB** field.

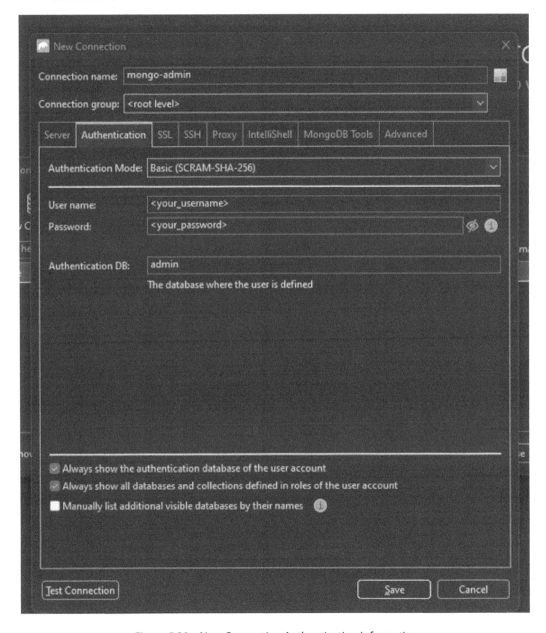

Figure 5.20 – New Connection Authentication information

3. **Testing our connection**: With this configuration, we should be able to test our database connection. In the lower-left corner, select the **Test Connection** button.

    If the credentials you provided are correct, you will see the following output:

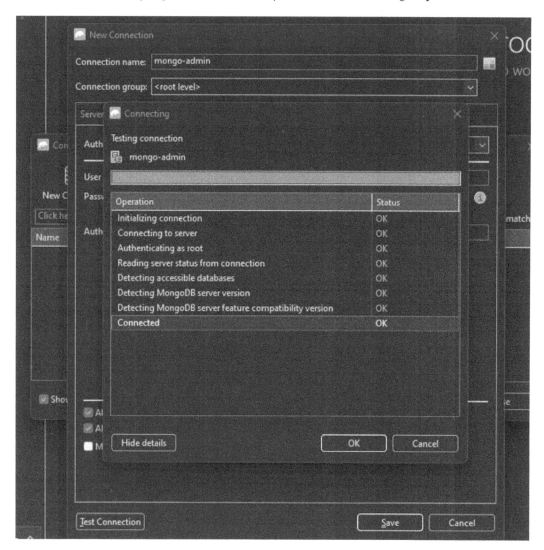

Figure 5.21 – Test connection successful

Click on the **Save** button, and the window will close.

4. **Connecting to our database**: After we save our configuration, a window with the available connections will appear, including our newly created one:

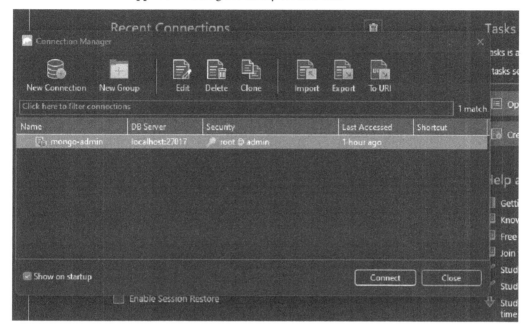

Figure 5.22 – Connection manager with the connection created

Select the **Connect** button, and three default databases will appear: **admin**, **config**, and **local**, as shown in the following screenshot:

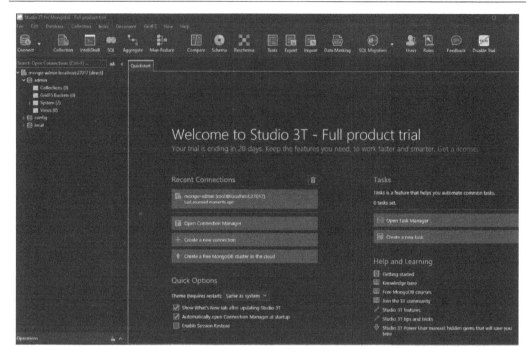

Figure 5.23 – The main page of the local MongoDB with the default databases on the server

We have now finished our MongoDB configuration and are ready for the following recipes in this chapter and others, including *Chapters 6*, *11*, and *12*.

How it works...

Like available databases, creating and running MongoDB through a Docker container is straightforward. Check the following commands:

```
my-project/mongo-local$ docker run \
--name mongodb-local \
-p 27017:27017 \
-e MONGO_INITDB_ROOT_USERNAME=<your_username> \
-e MONGO_INITDB_ROOT_PASSWORD=<your_password>\
-d mongo:latest
```

As we saw in the *Getting ready* section, the most crucial information to be passed is the username and password (using the -e parameter), the ports over which to connect (using the -p parameter), and the container image version, which is the latest available.

The architecture of the MongoDB container connected to Studio 3T Free is even more straightforward. Once the connection port is available, we can easily access the database. You can see the architectural representation as follows:

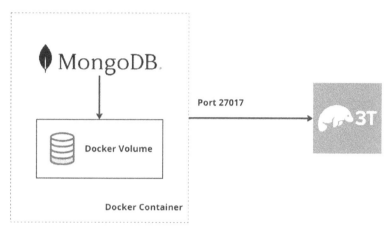

Figure 5.24 – MongoDB with Docker image connected to Studio 3T Free

As described at the beginning of this recipe, MongoDB is a document-oriented database. Its structure is similar to a JSON file, except each line is interpreted as a document and has its own `ObjectId`, as follows:

```
{
  "_id": "5cf0029caff5056591b0ce7d",
  "firstname": "Jane",
  "lastname": "Wu",
  "address": {
    "street": "1 Circle Rd",
    "city": "Los Angeles",
    "state": "CA",
    "zip": "90404"
  }
  "hobbies": ["surfing", "coding"]
}
```

Figure 5.25 – MongoDB document format

The group of documents is referred to as a *collection*, which is better understood as a table representation in a structured database. You can see how it is hierarchically organized in the schema shown here:

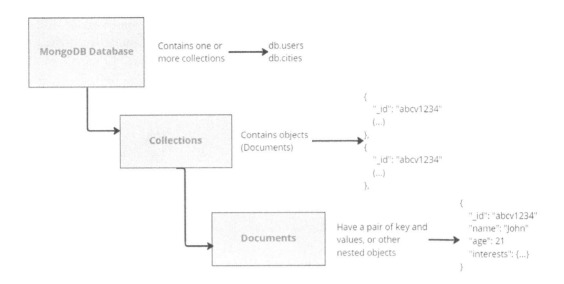

Figure 5.26 – MongoDB data structure

As we observed when logging in to MongoDB using Studio 3T Free, there are three default databases: admin, config, and local. For now, let's disregard the last two since they pertain to operational working of the data. The admin database is the main one created by the root user. That's why we provided this database for the **Authentication DB** option in *step 3*.

Creating a user to ingest data and access specific databases or collections is generally recommended. However, we will keep using root access here and in the following recipes in this book for demonstration purposes.

There's more...

The connection string will vary depending on how your MongoDB server is configured. For instance, when *replicas* or *sharded clusters* are in place, we need to specify which instances we want to connect to.

> **Note**
>
> Sharded clusters are a complex and interesting topic. You can read more and go deeper on the topic in MongoDB's official documentation at https://www.mongodb.com/docs/manual/core/sharded-cluster-components/.

Let's see an example of a standalone server string connection using basic authentication mode:

```
mongodb://mongo-server-user:some_password@mongo-host01.example.
com:27017/?authSource=admin
```

As you can see, it is similar to other database connections. If we wanted to connect to a local server, we would change the host to `localhost`.

Now, for a replica or sharded cluster, the string connection looks like this:

```
mongodb://mongo-server-user:some_password@mongo-host01.example.
com:27017, mongo-host02.example.com:27017, mongo-hosta03.example.
com:27017/?authSource=admin
```

The `authSource=admin` parameter in this URI is essential to inform MongoDB that we want to authenticate using the administration user of the database. Without it, an error or authentication will be raised, like the following output:

```
MongoError: Authentication failed
```

Another way to avoid this error is to create a specific user to access the database and collection.

SRV URI connection

MongoDB introduced the **Domain Name System** (**DNS**) seed list connection, constructed by a **DNS Service Record** (**SRV**) specification of data in the DNS, to try to solve this verbose string. We saw the possibility of using an SRV URI to configure the MongoDB connection in the first step of this recipe.

Here's an example of how it looks:

```
mongodb+srv://my-server.example.com/
```

It is similar to the standard connection string format we saw earlier. However, we need to indicate the use of SRV at the beginning and then provide the DNS entry.

This type of connection is advantageous when handling replicas or nodes since the SRV creates a single identity for the cluster. You can find a more detailed explanation of this, along with an outline of how to configure it, in the MongoDB official documentation at `https://www.mongodb.com/docs/manual/reference/connection-string/#dns-seed-list-connection-format`.

See also

If you are interested, other MongoDB GUI tools are available on the market: `https://www.guru99.com/top-20-mongodb-tools.html`.

Creating our NoSQL table in MongoDB

After successfully connecting and understanding how Studio 3T works, we will now import some MongoDB collections. We have seen in the *Connecting to a NoSQL database (MongoDB)* recipe how to get started with MongoDB, and in this recipe, we will import a MongoDB database and come to understand its structure. Although MongoDB has a specific format to organize data internally, understanding how a NoSQL database behaves is crucial when working with data ingestion.

We will practice by ingesting the imported collections in the following recipes in this chapter.

Getting ready

For this recipe, we will use a sample dataset of Airbnb reviews called listingsAndReviews.json. You can find this dataset in the GitHub repository of this book at https://github.com/PacktPublishing/Data-Ingestion-with-Python-Cookbook/tree/main/Chapter_5/datasets/sample_airbnb. After downloading it, put the file into our mongo-local directory, created in *Chapter 1*.

I kept mine inside the sample_airbnb folder just for organization purposes, as you can see in the following screenshot:

```
glauesppen@DESKTOP-DVUDB98:~/mongo-local/mongodb-sample-dataset/sample_airbnb$ ls
listingsAndReviews.json
```

Figure 5.27 – Command line with listingsAndReviews.json

After downloading the dataset, we need to install pymongo, a Python library to connect to and manage MongoDB operations. To install it, use the following command:

```
$ pip3 install pymongo
```

Feel free to create virtualenv for this installation.

We are now ready to start inserting data into MongoDB. Don't forget to check that your Docker image is up and running before we begin.

How to do it...

Here are the steps to perform this recipe:

1. **Creating our connection**: By importing and using pymongo, we can easily establish a connection with the MongoDB database. Refer to the following code:

    ```
    import json
    import os
    from pymongo import MongoClient, InsertOne
    mongo_client = pymongo.MongoClient("mongodb://root:root@
    localhost:27017/")
    ```

2. **Defining our database and collection**: We will create a database and collection instance using the client connection we instantiated.

For the `json_collection` variable, insert the path where you put the Airbnb sample dataset:

```
db_cookbook = mongo_client.db_airbnb
collection = db_cookbook.reviews
json_collection = "sample_airbnb/listingsAndReviews.json"
```

3. **Reading and bulk-inserting data**: Using the `bulk_write` function, we will insert all the documents inside the JSON file into the sales collection we created and close the connection:

```
requesting_collection = []
with open(json_collection) as f:
    for object in f:
        my_dict = json.loads(object)
        requesting.append(InsertOne(my_dict))

result = collection.bulk_write(requesting_collection)
mongo_client.close()
```

No output is expected from this operation, but we can check the database to see if it is successful.

4. **Checking the MongoDB database results**: Let's check our database to see if the data was inserted correctly.

Open Studio 3T Free and refresh the connection (right-click on the connection name and select **Refresh All**). You should see a new database named **db_airbnb** has been created, containing a **reviews** collection, as shown in the following screenshot:

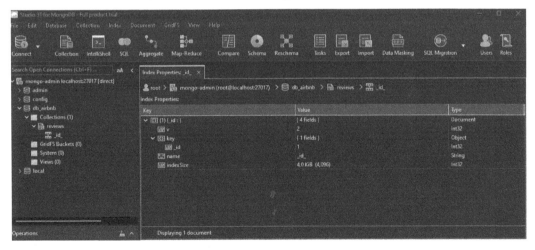

Figure 5.28 – Database and collection successfully created on MongoDB

With the collection now created and containing some data, let's go deeper into how the code works.

How it works...

As you can see, the code we implemented is straightforward, using just a few lines to create and insert data in our database. However, there are important points to pay attention to due to the particularities of MongoDB.

Let's examine the code line by line now:

```
mongo_client = pymongo.MongoClient("mongodb://root:root@
localhost:27017/")
```

This line defines the connection to our MongoDB database, and from this instance, we can create a new database and its collections.

> **Note**
>
> Observe that the URI connection contains hardcoded values for the username and password. This must be avoided in real applications, and even development servers. It is recommended to store those values as environment variables or use a secret manager vault.

Next, we define the database and collection names; you may have noticed we didn't create them previously in our database. At the time of execution of the code, MongoDB checks whether the database exists; if not, MongoDB will create it. The same rule applies to the **reviews** collection.

Notice the collection derives from the db_cookbook instance, which makes it clear that it is linked to the db_airbnb database:

```
db_cookbook = mongo_client.db_airbnb
collection = db_cookbook.reviews
```

Following the code, the next step is to open the JSON file and parse every line. Here we start to see some tricky peculiarities of MongoDB:

```
requesting_collection = []
with open(json_collection) as f:
    for object in f:
        my_dict = json.loads(object)
        requesting_collection.append(InsertOne(my_dict))
```

It is common to wonder why we actually need to parse the lines of JSON, since MongoDB accepts this format. Let's check our listingsAndReviews.json file, as shown in the following screenshot:

Figure 5.29 – JSON file with MongoDB document lines

If we use any tool to verify this as valid JSON, it will certainly say it's not a valid format. This happens because each line of this file represents one document of the MongoDB collection. Trying to open that file using only the conventional `open()` and `json.loads()` methods will produce an error like the following:

```
json.decoder.JSONDecodeError: Extra data: line 2 column 1 (char 190)
```

To make it acceptable to the Python interpreter, we need to open and read each line individually and append it to the `requesting_collection` list. Also, the `InsertOne()` method will ensure that each line is inserted separately. A problem that occurs while inserting a specific row will be much easier to identify.

Finally, the `bulk_write()` will take the list of documents and insert them into the MongoDB database:

```
result = collection.bulk_write(requesting_collection)
```

This operation will finish without returning any output or error messages if everything is OK.

There's more…

We have seen how simple it is to create a Python script to insert data into our MongoDB server. Nevertheless, MongoDB has database tools to provide the same result and can be executed via the command line. The `mongoimport` command is used to insert data into our database, as you can see in the following code:

```
mongoimport --host localhost --port 27017-d db_name -c collection_name
--file path/to/file.json
```

If you are interested to learn more about the other database tools and commands available, check the official MongoDB documentation at `https://www.mongodb.com/docs/database-tools/installation/installation/`.

Restrictions on field names

When loading data into MongoDB, one big problem is the restrictions on characters used in the field names. Due to MongoDB server versions or programming language specificities, sometimes the key names of fields come with a $ prefix, and, by default, MongoDB is not compatible with it, creating an error like the following output:

```
localhost:27017: $oid is not valid for storage.
```

In this case, a JSON file dump was exported from a MongoDB server, and the reference of `ObjectID` came with the $ prefix. Even though the more recent versions of MongoDB have started to accept these characters (see the thread here: `https://jira.mongodb.org/browse/SERVER-41628?fbcl id=IwAR1t5Ld58LwCi69SrMCcDbhPGf2EfBWe_AEurxGkEWHpZTHaEIde0_AZ-uM%5D`), it is a good practice to avoid using them where possible. In this case, we have two main options: remove all the restricted characters using a script, or encode the JSON file into a **Binary JavaScript Object Notation** (**BSON**) file. You can find out more about encoding the file into BSON format at `https://kb.objectrocket.com/mongo-db/how-to-use-python-to-encode-a-json-file-into-mongodb-bson-documents-545`.

See also

You can read more about the MongoDB restrictions on field names at `https://www.mongodb.com/docs/manual/reference/limits/#mongodb-limit-Restrictions-on-Field-Names`.

Ingesting data from MongoDB using PySpark

Even though it seems impractical to create and ingest the data ourselves, this exercise can be applied to real-life projects. People who work with data are often involved in the architectural process of defining the type of database, helping other engineers to insert data from applications into a database server, and later ingesting only the relevant information for dashboards or other analytical tools.

So far, we have created and evaluated our server and then created collections inside our MongoDB instance. With all this preparation, we can now ingest our data using PySpark.

Getting ready

This recipe requires the execution of the *Creating our NoSQL table in MongoDB* recipe due to data insertion. However, you can create and insert other documents into the MongoDB database and use them here. If you do this, ensure you set the suitable configurations to make it run properly.

Also, as in the *Creating our NoSQL table in MongoDB* recipe, check that the Docker container is up and running since this is our MongoDB instance's primary data source. Let's proceed to the ingesting!

Figure 5.30 – Docker container for MongoDB is running

How to do it...

You need to perform the following steps to try this recipe:

1. **Creating the SparkSession**: As usual, the first thing to do is create `SparkSession`, but this time passing specific configurations for reading our MongoDB database, `db_airbnb`, such as the URI and the `.jars`:

    ```
    from pyspark.sql import SparkSession
    spark = SparkSession.builder \
            .master("local[1]") \
            .appName("MongoDB Ingest") \
            .config("spark.executor.memory", '3g') \
            .config("spark.executor.cores", '1') \
            .config("spark.cores.max", '1') \
            .config("spark.mongodb.input.
    uri", "mongodb://root:root@127.0.0.1/db_
    airbnb?authSource=admin&readPreference=primaryPreferred") \
            .config("spark.mongodb.input.collection", "reviews") \
            .config("spark.jars.packages","org.mongodb.spark:mongo-
    spark-connector_2.12:3.0.1") \
            .getOrCreate()
    ```

We should expect a significant output here since Spark downloads the package and sets the rest of the configuration we passed:

```
:: loading settings :: url = jar:file:/home/glauesppen/.local/lib/python3.8/site-packages/pyspark/jars/ivy-2.4.0.jar!/org/apach
e/ivy/core/settings/ivysettings.xml

Ivy Default Cache set to: /home/glauesppen/.ivy2/cache
The jars for the packages stored in: /home/glauesppen/.ivy2/jars
org.mongodb.spark#mongo-spark-connector_2.12 added as a dependency
:: resolving dependencies :: org.apache.spark#spark-submit-parent-090c6cef-ae6e-4343-b164-71661911f14d;1.0
        confs: [default]
        found org.mongodb.spark#mongo-spark-connector_2.12;3.0.1 in central
        found org.mongodb#mongodb-driver-sync;4.0.5 in central
        found org.mongodb#bson;4.0.5 in central
        found org.mongodb#mongodb-driver-core;4.0.5 in central
:: resolution report :: resolve 90ms :: artifacts dl 2ms
        :: modules in use:
        org.mongodb#bson;4.0.5 from central in [default]
        org.mongodb#mongodb-driver-core;4.0.5 from central in [default]
        org.mongodb#mongodb-driver-sync;4.0.5 from central in [default]
        org.mongodb.spark#mongo-spark-connector_2.12;3.0.1 from central in [default]
        ---------------------------------------------------------------------
        |                  |            modules            ||   artifacts   |
        |       conf       | number| search|dwnlded|evicted|| number|dwnlded|
        ---------------------------------------------------------------------
        |     default      |   4   |   0   |   0   |   0   ||   4   |   0   |
        ---------------------------------------------------------------------
:: retrieving :: org.apache.spark#spark-submit-parent-090c6cef-ae6e-4343-b164-71661911f14d
        confs: [default]
        0 artifacts copied, 4 already retrieved (0kB/3ms)
22/12/15 22:27:24 WARN NativeCodeLoader: Unable to load native-hadoop library for your platform... using builtin-java classes w
here applicable
Using Spark's default log4j profile: org/apache/spark/log4j-defaults.properties
Setting default log level to "WARN".
To adjust logging level use sc.setLogLevel(newLevel). For SparkR, use setLogLevel(newLevel).
```

Figure 5.31 – SparkSession being initialized with MongoDB configurations

1. **Reading the reviews collection**: With the connection established, we can now read the collection using `SparkSession` we instantiated. Here, no output is expected because the `SparkSession` is set only to send logs at the `WARN` level:

    ```
    df = spark.read.format("mongo").load()
    ```

2. **Getting our DataFrame schema**: We can see the collection's schema using the print operation on the DataFrame:

    ```
    df.printSchema()
    ```

You should observe the following output:

```
root
 |-- _id: string (nullable = true)
 |-- access: string (nullable = true)
 |-- accommodates: struct (nullable = true)
 |    |-- $numberInt: string (nullable = true)
 |-- address: struct (nullable = true)
 |    |-- street: string (nullable = true)
 |    |-- suburb: string (nullable = true)
 |    |-- government_area: string (nullable = true)
 |    |-- market: string (nullable = true)
 |    |-- country: string (nullable = true)
 |    |-- country_code: string (nullable = true)
 |    |-- location: struct (nullable = true)
 |    |    |-- type: string (nullable = true)
 |    |    |-- coordinates: array (nullable = true)
 |    |    |    |-- element: struct (containsNull = true)
 |    |    |    |    |-- $numberDouble: string (nullable = true)
 |    |    |-- is_location_exact: boolean (nullable = true)
 |-- amenities: array (nullable = true)
 |    |-- element: string (containsNull = true)
 |-- availability: struct (nullable = true)
 |    |-- availability_30: struct (nullable = true)
 |    |    |-- $numberInt: string (nullable = true)
 |    |-- availability_60: struct (nullable = true)
 |    |    |-- $numberInt: string (nullable = true)
 |    |-- availability_90: struct (nullable = true)
 |    |    |-- $numberInt: string (nullable = true)
 |    |-- availability_365: struct (nullable = true)
 |    |    |-- $numberInt: string (nullable = true)
 |-- bathrooms: struct (nullable = true)
 |    |-- $numberDecimal: string (nullable = true)
 |-- bed_type: string (nullable = true)
 |-- bedrooms: struct (nullable = true)
 |    |-- $numberInt: string (nullable = true)
 |-- beds: struct (nullable = true)
```

Figure 5.32 – Reviews DataFrame collection schema printed

As you can observe, the structure is similar to a JSON file with nested objects. Unstructured data is usually presented in this form and can hold a large amount of information to create a Python script to insert data into our data. Now, let's go deeper and understand our code.

How it works...

MongoDB required a few additional configurations in `SparkSession` to execute the `.read` function. It is essential to understand why we used the configurations instead of just using code from the documentation. Let's explore the code for it:

```
config("spark.mongodb.input.uri", "mongodb://root:root@127.0.0.1/db_
airbnb?authSource=admin) \
```

Note the use of `spark.mongodb.input.uri`, which tells our `SparkSession` that a *read* operation needs to be performed using a MongoDB URI. If, for instance, we wanted to do a *write* operation (or both read and write), we would just need to add the `spark.mongodb.output.uri` configuration.

Next, we pass the URI containing the user and password information, the name of the database, and the authentication source. Since we use the root user to retrieve the data, this last parameter is set to `admin`.

Next, we define the name of our collection to be used in the read operation:

```
.config("spark.mongodb.input.collection", "reviews")\
```

> **Note**
>
> Even though it might seem odd to define these parameters in the SparkSession, and it is possible to set the database and collection, this is a good practice that has been adopted by the community when manipulating MongoDB connections.

```
.config("spark.jars.packages","org.mongodb.spark:mongo-spark-connector_2.12:3.0.1")
```

Another new configuration here is `spark.jars.packages`. When using this key with the `.config()` method, Spark will search its available online packages, download them, and place them in the `.jar` folders to be used. Although this is an advantageous way to set the `.jar` connectors, this is not available for all databases.

Once the connection is established, the reading process is remarkably similar to the JDBC: we pass the `.format()` of the database (here, `mongo`), and since the database and collection name are already set, we don't need to configure `.option()`:

```
df = spark.read.format("mongo").load()
```

When executing `.load()`, Spark will verify whether the connection is valid and throw an error if not. In the following screenshot, you can see an example of the error message when the credentials are not correct:

```
Py4JJavaError: An error occurred while calling o57.load.
: com.mongodb.MongoSecurityException: Exception authenticating MongoCredential{mechanism=SCRAM-SHA-256, userName='root', source
='admin', password=<hidden>, mechanismProperties=<hidden>}
        at com.mongodb.internal.connection.SaslAuthenticator.wrapException(SaslAuthenticator.java:201)
        at com.mongodb.internal.connection.SaslAuthenticator.access$300(SaslAuthenticator.java:40)
        at com.mongodb.internal.connection.SaslAuthenticator$1.run(SaslAuthenticator.java:78)
        at com.mongodb.internal.connection.SaslAuthenticator$1.run(SaslAuthenticator.java:47)
        at com.mongodb.internal.connection.SaslAuthenticator.doAsSubject(SaslAuthenticator.java:207)
        at com.mongodb.internal.connection.SaslAuthenticator.authenticate(SaslAuthenticator.java:47)
        at com.mongodb.internal.connection.InternalStreamConnectionInitializer.authenticate(InternalStreamConnectionInitialize
r.java:152)
```

Figure 5.33 – Py4JJavaError: Authentication error to MongoDB connection

Even though we are handling an unstructured data format, as soon as PySpark transforms our collection into a DataFrame, all the filtering, cleaning, and manipulating of data is pretty much the same as PySpark data.

There's more...

As we saw previously, PySpark error messages can be confusing and cause discomfort at first glance. Let's explore other common errors when ingesting data from a MongoDB database without the proper configuration.

In this example, let's not set `spark.jars.packages` in the `SparkSession` configuration:

```
spark = SparkSession.builder \
    (...)
    .config("spark.mongodb.input.uri", "mongodb://
root:root@127.0.0.1/db_aibnb?authSource=admin") \
    .config("spark.mongodb.input.collection", "reviews")
    .getOrCreate()

df = spark.read.format("mongo").load()
```

If you try to execute the preceding code (passing the rest of the memory settings), you will get the following output:

```
~/.local/lib/python3.8/site-packages/py4j/protocol.py in get_return_value(answer, gateway_client, target_id, name)
    324            value = OUTPUT_CONVERTER[type](answer[2:], gateway_client)
    325            if answer[1] == REFERENCE_TYPE:
--> 326                raise Py4JJavaError(
    327                    "An error occurred while calling {0}{1}{2}.\n".
    328                    format(target_id, ".", name), value)

Py4JJavaError: An error occurred while calling o50.load.
: java.lang.ClassNotFoundException: Failed to find data source: mongo. Please find packages at http://spark.apache.org/third-pa
rty-projects.html
        at org.apache.spark.sql.execution.datasources.DataSource$.lookupDataSource(DataSource.scala:692)
        at org.apache.spark.sql.execution.datasources.DataSource$.lookupDataSourceV2(DataSource.scala:746)
        at org.apache.spark.sql.DataFrameReader.load(DataFrameReader.scala:265)
        at org.apache.spark.sql.DataFrameReader.load(DataFrameReader.scala:225)
        at sun.reflect.NativeMethodAccessorImpl.invoke0(Native Method)
        at sun.reflect.NativeMethodAccessorImpl.invoke(NativeMethodAccessorImpl.java:62)
        at sun.reflect.DelegatingMethodAccessorImpl.invoke(DelegatingMethodAccessorImpl.java:43)
        at java.lang.reflect.Method.invoke(Method.java:498)
        at py4j.reflection.MethodInvoker.invoke(MethodInvoker.java:244)
        at py4j.reflection.ReflectionEngine.invoke(ReflectionEngine.java:357)
        at py4j.Gateway.invoke(Gateway.java:282)
        at py4j.commands.AbstractCommand.invokeMethod(AbstractCommand.java:132)
        at py4j.commands.CallCommand.execute(CallCommand.java:79)
        at py4j.GatewayConnection.run(GatewayConnection.java:238)
        at java.lang.Thread.run(Thread.java:750)
Caused by: java.lang.ClassNotFoundException: mongo.DefaultSource
        at java.net.URLClassLoader.findClass(URLClassLoader.java:387)
        at java.lang.ClassLoader.loadClass(ClassLoader.java:418)
        at java.lang.ClassLoader.loadClass(ClassLoader.java:351)
        at org.apache.spark.sql.execution.datasources.DataSource$.$anonfun$lookupDataSource$5(DataSource.scala:666)
        at scala.util.Try$.apply(Try.scala:213)
        at org.apache.spark.sql.execution.datasources.DataSource$.$anonfun$lookupDataSource$4(DataSource.scala:666)
        at scala.util.Failure.orElse(Try.scala:224)
        at org.apache.spark.sql.execution.datasources.DataSource$.lookupDataSource(DataSource.scala:666)
        ... 14 more
```

Figure 5.34 – java.lang.ClassNotFoundException error when the
MongoDB package is not set in the configuration

Looking carefully at the second line, which begins with `java.lang.ClassNotFoundException`, the JVM highlights a missing package or class that needs to be searched for in a third-party repository. The package contains the connector code to our JVM and establishes communication with the database server.

Another widespread error message is `IllegalArgumentException`. This type of error indicates to the developer that an argument was wrongly passed to a method or class. Usually, when related to database connections, it refers to an invalid string connection, as in the following screenshot:

```
~/.local/lib/python3.8/site-packages/pyspark/sql/utils.py in deco(*a, **kw)
    115                 # Hide where the exception came from that shows a non-Pythonic
    116                 # JVM exception message.
--> 117                 raise converted from None
    118             else:
    119                 raise

IllegalArgumentException: requirement failed: Invalid uri: 'mongodb://root:roo@127.0.0.1/db_aibnb/?authSource=admin&readPreference=primaryPreferred'
```

Figure 5.35 – IllegalArgumentException error when the URI is invalid

Although it seems unclear, there is a typo in the URI, where `db_aibnb/?` contains an extra forward slash. Removing it and running `SparkSession` again will make this error disappear.

> **Note**
> It is recommended to shut down and restart the kernel processes when re-defining the SparkSession configurations because SparkSession tends to append to the processes rather than replacing them.

See also

- MongoDB Spark connector documentation: `https://www.mongodb.com/docs/spark-connector/current/configuration/`

- You can check the MongoDB documentation for a full explanation of how the MongoDB connector behaves with PySpark: `https://www.mongodb.com/docs/spark-connector/current/read-from-mongodb/`

- There are also some interesting use cases of MongoDB here: `https://www.mongodb.com/use-cases`

Further reading

- https://www.talend.com/resources/structured-vs-unstructured-data/
- https://careerfoundry.com/en/blog/data-analytics/structured-vs-unstructured-data/
- https://www.dba-ninja.com/2022/04/is-mongodbsrv-necessary-for-a-mongodb-connection.html
- https://www.mongodb.com/docs/manual/reference/connection-string/#connection-string-options
- https://sparkbyexamples.com/spark/spark-createorreplacetempview-explained/

Using PySpark with Defined and Non-Defined Schemas

Generally, schemas are forms used to create or apply structures to data. As someone who works or will work with large volumes of data, it is essential to understand how to manipulate DataFrames and apply structure when it is necessary to bring more context to the information involved.

However, as seen in the previous chapters, data can come from different sources or be present without a well-defined structure, and applying a schema can be challenging. Here, we will see how to create schemas and standard formats using PySpark with structured and unstructured data.

In this chapter, we will cover the following recipes:

- Applying schemas to data ingestion
- Importing structured data using a well-defined schema
- Importing unstructured data with an undefined schema
- Ingesting unstructured data with a well-defined schema and format
- Inserting formatted SparkSession logs to facilitate your work

Technical requirements

You can also find the code for this chapter in the GitHub repository here: `https://github.com/PacktPublishing/Data-Ingestion-with-Python-Cookbook`.

Using **Jupyter Notebook** is not mandatory but can help you see how the code works interactively. Since we will execute Python and PySpark code, it can help us understand the scripts better. Once you have it installed, you can execute Jupyter using the following line:

```
$ jupyter Notebook
```

It is recommended to create a separate folder to store the Python files or Notebooks we will cover in this chapter; however, feel free to organize the files in the best way that fits you.

In this chapter, all recipes will need a `SparkSession` instance initialized, and you can use the same session for all of them. You can use the following code to create your session:

```python
from pyspark.sql import SparkSession
spark = SparkSession.builder \
        .master("local[1]") \
        .appName("chapter6_schemas") \
        .config("spark.executor.memory", '3g') \
        .config("spark.executor.cores", '1') \
        .config("spark.cores.max", '1') \
        .getOrCreate()
```

> **Note**
> A WARN message as output is expected in some cases, especially if you are using WSL on Windows, so you don't need to worry.

Applying schemas to data ingestion

The application of schemas is common practice when ingesting data, and PySpark natively supports applying them to DataFrames. To define and apply schemas to our DataFrames, we need to understand some concepts of Spark.

This recipe introduces the basic concept of working with schemas using PySpark and its best practices so that we can later apply them to structured and unstructured data.

Getting ready

Make sure PySpark is installed and working on your machine for this recipe. You can run the following code on your command line to check this requirement:

```
$ pyspark --version
```

You should see the following output:

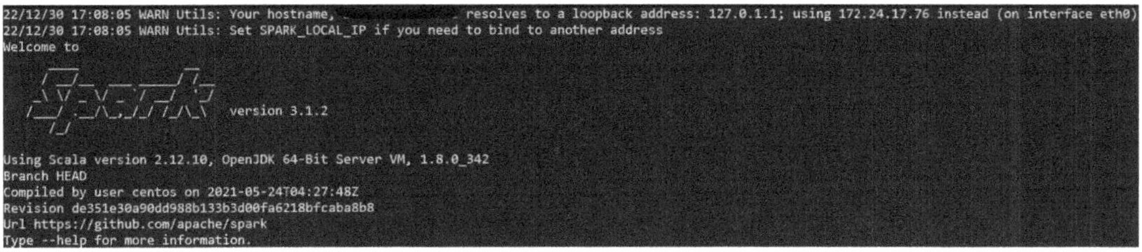

Figure 6.1 – PySpark version console output

If don't have PySpark installed on your local machine, please refer to the *Installing PySpark* recipe in *Chapter 1*.

I will use Jupyter Notebook to execute the code to make it more interactive. You can use this link and follow the instructions on the screen to install it: `https://jupyter.org/install`.

If you already have it installed, check the version using the following code on your command line:

```
$ jupyter --version
```

The following screenshot shows the expected output:

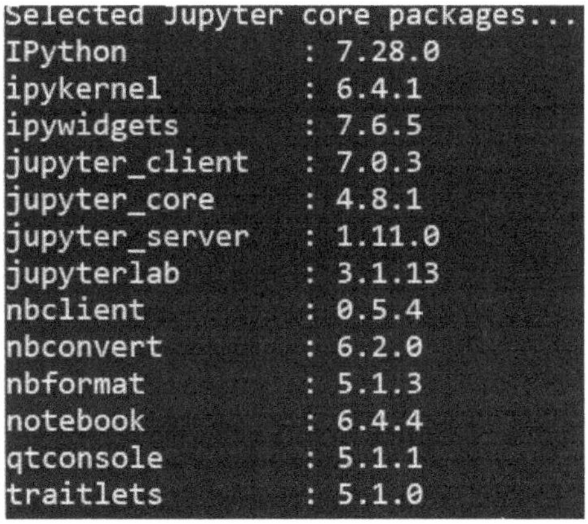

Figure 6.2 – Jupyter package versions

As you can see, the Notebook version was 6.4.4 at the time this book was written. Make sure to always use the latest version.

How to do it...

Here are the steps to carry out the recipe:

1. **Creating mock data**: Before applying the schema to our DataFrame, we need to create a simple set containing simulated data of people's information in this format—ID, name, last name, age, and gender:

    ```
    my_data = [("3456","Cristian","Rayner",30,"M"),
               ("3567","Guto","Flower",35,"M"),
               ("9867","Yasmin","Novak",23,"F"),
               ("3342","Tayla","Mejia",45,"F"),
               ("8890","Barbara","Kumar",20,"F")
               ]
    ```

2. **Importing and structuring the schema**: The next step is to import the types and create the structure of our schema:

    ```
    from pyspark.sql.types import StructType, StructField,
    StringType, IntegerType
    schema = StructType([ \
        StructField("id",StringType(),True), \
        StructField("name",StringType(),True), \
        StructField("lastname",StringType(),True), \
        StructField("age", IntegerType(), True), \
        StructField("gender", StringType(), True), \
      ])
    ```

3. **Creating the DataFrame**: Then, we make the DataFrame, applying the schema we have created:

    ```
    df = spark.createDataFrame(data=my_data,schema=schema)
    ```

    When printing our DataFrame schema using the .printSchema() method, this is the expected output:

    ```
    root
     |-- id: string (nullable = true)
     |-- name: string (nullable = true)
     |-- lastname: string (nullable = true)
     |-- age: integer (nullable = true)
     |-- gender: string (nullable = true)
    ```

    Figure 6.3 – The DataFrame schema

How it works...

Before understanding the methods in *step 2*, let's step back a bit and understand the concept of a DataFrame.

DataFrame is like a table with data stored and organized in a two-dimensional array, which resembles a table from a relational database such as MySQL or Postgres. Each line corresponds to a record, and libraries such as pandas and PySpark, by default, assign a record number internally to each line (or index).

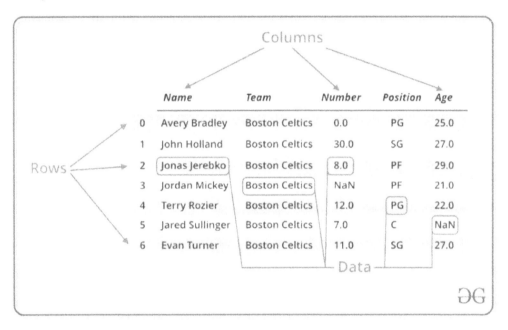

Figure 6.4 – GeeksforGeeks DataFrame explanation

> **Note**
>
> Speaking of the Pandas library, it is common to refer to a column in a Pandas DataFrame as a series and expect it to behave like a Python list. It makes it easier to manipulate data for analysis and visualization.

The objective of using a DataFrame is to utilize the several optimizations for data processing under the hood that Spark brings, which are directly linked to parallel processing.

Back to the schema definition in our `schema` variable, let's take a look at the code we created:

```
schema = StructType([ \
    StructField("id",StringType(),True), \
```

```
    StructField("name",StringType(),True), \
    StructField("lastname",StringType(),True), \
    StructField("age", IntegerType(), True), \
    StructField("gender", StringType(), True), \
])
```

The first object we declare is the `StructType` class. This class will create an object of collection or our rows. Next, we declare a `StructField` instance, representing our column with its name, data type, whether it is nullable, and its metadata when applicable. `StructField` must be in the same order as the columns in the DataFrame; otherwise, it can generate errors due to the incompatibility of a data type (for example, the column has string values, and we are setting it as an integer) or the presence of null values. Defining `StructField` is an excellent opportunity to standardize the name of the DataFrame and, therefore, the analytical data.

Finally, `StringType` and `IntegerType` are the methods that will cast the data type into the respective columns. They were imported at the beginning of the recipe and derived from the SQL types inside PySpark. In our mock data example, we defined the `id` and `age` columns as `IntegerType` since we expect no other kind of data in them. However, there are many situations where the `id` column is referred to as a string type, usually when the data comes from different systems.

Here we used `StringType` and `IntegerType`, but many others can be used to create context and standardize our data. Refer to the following figure:

StringType	ShortType
ArrayType	IntegerType
MapType	LongType
StructType	FloatType
DateType	DoubleType
TimestampType	DecimalType
BooleanType	ByteType
CalendarIntervalType	HiveStringType
BinaryType	ObjectType
NumericType	NullType

Figure 6.5 – SparkbyExample table of data types in Spark

You can understand more about how to apply the **Spark data types** in the Spark official documentation here: `https://spark.apache.org/docs/latest/sql-ref-datatypes.html`.

There's more...

When handling terminology, there needs to be a common understanding about using a dataset or DataFrame, especially if you are a newcomer to the data world.

A dataset is a collection of data containing rows and columns (for example, relational data) or documents and files (for example, non-relational data). It comes from a source and is available in different file formats.

On the other hand, a DataFrame is derived from a dataset, presenting the data in a tabular form even if that is not the primary format. The DataFrame can transform a MongoDB document collection into a tabular organization based on the configurations set when it is created.

See also

More examples on the *SparkbyExample* site can be found here: `https://sparkbyexamples.com/pyspark/pyspark-sql-types-datatype-with-examples/`.

Importing structured data using a well-defined schema

As seen in the previous chapter, *Ingesting Data from Structured and Unstructured Databases*, structured data has a standard format presented in rows and columns and is often stored inside a database.

Due to its format, the application of a DataFrame schema tends to be less complex and has several benefits, such as ensuring the ingested information is the same as the data source or follows a rule.

In this recipe, we will ingest data from a structured file such as a CSV file and apply a DataFrame schema to understand better how it is used in a real-world scenario.

Getting ready

This exercise requires the `listings.csv` file found inside the GitHub repository for this book. Also, make sure your `SparkSession` is initialized.

All the code in this recipe can be executed in Jupyter Notebook cells or a PySpark shell.

How to do it...

Here are the steps to perform this recipe:

1. **Importing Spark data types**: Besides `StringType` and `IntegerType`, we will include two more data types in our import, `FloatType` and `DateType`, shown as follows:

    ```
    from pyspark.sql.types import StructType, StructField,
    StringType, IntegerType, FloatType, DateType, DoubleType
    ```

2. **Creating the schema**: Based on the columns inside the CSV file, let's insert their names into each `StructField` and assign them to a respective data type:

```
schema = StructType([ \
    StructField("id",IntegerType(),True), \
    StructField("name",StringType(),True), \
    StructField("host_id",IntegerType(),True), \
    StructField("host_name",StringType(),True), \
    StructField("neighbourhood_group",StringType(),True), \
    StructField("neighbourhood",StringType(),True), \
    StructField("latitude",DoubleType(),True), \
    StructField("longitude",DoubleType(),True), \
    StructField("room_type",StringType(),True), \
    StructField("price",FloatType(),True), \
    StructField("minimum_nights",IntegerType(),True), \
    StructField("number_of_reviews",IntegerType(),True), \
    StructField("last_review",DateType(),True), \
    StructField("reviews_per_month",FloatType(),True), \
        StructField("calculated_host_listings_
count",IntegerType(),True), \
    StructField("availability_365",IntegerType(),True), \
    StructField("number_of_reviews_ltm",IntegerType(),True), \
    StructField("license",StringType(),True)
  ])
```

3. **Reading the CSV file with the schema defined**: Now let's read our `listings.csv` file with the `.options()` configurations and add the `.schema()` method with the `schema` variable seen in *step 2*.

```
df = spark.read.options(header=True, sep=',',
                        multiLine=True, escape='"')\
            .schema(schema) \
            .csv('listings.csv')
```

If everything is well set, you should see no output from this execution.

4. **Checking the read DataFrame**: We can check the schema of our DataFrame by executing the following code:

```
df.printSchema()
```

You should see the following output:

```
root
 |-- id: integer (nullable = true)
 |-- name: string (nullable = true)
 |-- host_id: integer (nullable = true)
 |-- host_name: string (nullable = true)
 |-- neighbourhood_group: string (nullable = true)
 |-- neighbourhood: string (nullable = true)
 |-- latitude: double (nullable = true)
 |-- longitude: double (nullable = true)
 |-- room_type: string (nullable = true)
 |-- price: float (nullable = true)
 |-- minimum_nights: integer (nullable = true)
 |-- number_of_reviews: integer (nullable = true)
 |-- last_review: date (nullable = true)
 |-- reviews_per_month: float (nullable = true)
 |-- calculated_host_listings_count: integer (nullable = true)
 |-- availability_365: integer (nullable = true)
 |-- number_of_reviews_ltm: integer (nullable = true)
 |-- license: string (nullable = true)
```

Figure 6.6 – listings.csv DataFrame schema

How it works...

We made a few additions in this exercise that differ from the last recipe, *Applying schemas to data ingestion*.

As usual, we started by importing the required methods to make the script work. We added three more data types: float, double, and date. The choice was made based on what the CSV file contained. Let's look at the first lines of our file, as you can see in the following screenshot:

A	B	C	D	E	F	G	H	I	J	K	L	M	N	O	P	Q	R	S	T
id	name	host_id	host_nam	neighbou	neighbou	latitude	longitude	room_typ	price	minimum	number_c	last_review	reviews_p	calculatec	availabilit	number_c	license		
2818	Quiet Gar	3159	Daniel		Oostelijk I	5.236.435	494.358	Private ro	49	3	285	21/11/2021	1.81	1	62	7	0363 5F3A 5684 6750 D14D		
20168	Studio wit	59484	Alexander		Centrum-(	5.236.407	489.393	Private ro	106	1	339	09/04/2020	2.30	2	0	0	0363 CBB3 2C10 0C2A 1E29		
27886	Romantic,	97647	Flip		Centrum-'	5.238.761	489.188	Private ro	134	2	228	20/02/2022	1.84	1	189	9	0363 974D 4986 7411 88D8		
28871	Comfortal	124245	Edwin		Centrum-'	5.236.775	489.092	Private ro	75	2	379	05/03/2022	2.70	2	146	43	0363 607B EA74 0BD8 2F6F		
29051	Comfortal	124245	Edwin		Centrum-(	5.236.584	489.111	Private ro	55	2	532	04/03/2022	3.98	2	170	53	0363 607B EA74 0BD8 2F6F		
44391	Quiet 2-b	194779	Jan		Centrum-(	5.237.168	491.471	Entire hor	240	3	41	28/08/2021	0.29	1	0	2	0363 E76E F06A C1DD 172C		

Figure 6.7 – listings.csv view from Microsoft Excel

We can observe different types of numerical fields; some require more decimal places, and `last_review` is in a date format. Because of this, we made additional library imports, as you can see in the following piece of code:

```
from pyspark.sql.types import StructType, StructField, StringType,
IntegerType, FloatType, DoubleType, DateType
```

Step 2 is similar to what we did before, where we attributed the column names and their respective data types. It is in *step 3* that we made the schema attribution using the `schema()` method from the `SparkSession` class. If the schema contains the same number of columns as the file, we should expect no output here; otherwise, this message will appear:

```
22/12/30 22:21:45 WARN CSVHeaderChecker: Number of column in CSV header is not equal to number of fields in the schema:
 Header length: 18, schema size: 15
CSV file: file:///home/glauesppen/listings.csv
```

Figure 6.8 – Output warning message when the schema does not match

The content is important, even if it is a WARN log message. Looking closely, it says the schema does not match the number of columns in the file. This could be a problem later in the ETL pipeline when loading data into a data warehouse or any other analytical database.

There's more...

If you go back to *Chapter 4*, you will observe that one of our CSV readings contains an inferSchema parameter inserted into the options() method. Refer to the following code:

```
df_2 = spark.read.options(header=True, sep=',',
                        multiLine=True, escape='"',
                        inferSchema=True) \
            .csv('listings.csv')
```

This parameter tells Spark to infer the data types based on the traits of the rows. For example, if a row has a value without quotation marks and is a number, there is a good chance this is an integer. However, if it contains a quotation mark, Spark can interpret it as a string, and any numerical operation will break.

In the recipe, if we use inferSchema, we will see a very similar printSchema output from the schema we defined, except for some fields that were interpreted as DoubleType and that we declared as FloatType or DateType. This is shown in the following screenshot:

```
root
 |-- id: integer (nullable = true)
 |-- name: string (nullable = true)
 |-- host_id: integer (nullable = true)
 |-- host_name: string (nullable = true)
 |-- neighbourhood_group: string (nullable = true)
 |-- neighbourhood: string (nullable = true)
 |-- latitude: double (nullable = true)
 |-- longitude: double (nullable = true)
 |-- room_type: string (nullable = true)
 |-- price: float (nullable = true)
 |-- minimum_nights: integer (nullable = true)
 |-- number_of_reviews: integer (nullable = true)
 |-- last_review: date (nullable = true)
 |-- reviews_per_month: float (nullable = true)
 |-- calculated_host_listings_count: integer (nullable = true)
 |-- availability_365: integer (nullable = true)
 |-- number_of_reviews_ltm: integer (nullable = true)
 |-- license: string (nullable = true)
```

```
root
 |-- id: long (nullable = true)
 |-- name: string (nullable = true)
 |-- host_id: integer (nullable = true)
 |-- host_name: string (nullable = true)
 |-- neighbourhood_group: string (nullable = true)
 |-- neighbourhood: string (nullable = true)
 |-- latitude: double (nullable = true)
 |-- longitude: double (nullable = true)
 |-- room_type: string (nullable = true)
 |-- price: integer (nullable = true)
 |-- minimum_nights: integer (nullable = true)
 |-- number_of_reviews: integer (nullable = true)
 |-- last_review: string (nullable = true)
 |-- reviews_per_month: double (nullable = true)
 |-- calculated_host_listings_count: integer (nullable = true)
 |-- availability_365: integer (nullable = true)
 |-- number_of_reviews_ltm: integer (nullable = true)
 |-- license: string (nullable = true)
```

           With schema applied   Using inferSchema

Figure 6.9 – Schema comparison when using a schema with inferSchema

Even though it seems like a tiny detail, when dealing with streaming or a large dataset and few computational resources, it can make a difference. Float types have a small range and bring in high processing power. Double data types bring more precision and are used to decrease mathematical errors or values being rounded by decimal data types.

See also

Read more about float and double types on the *Hackr IO* website here: `https://hackr.io/blog/float-vs-double`.

Importing unstructured data without a schema

As seen before, unstructured data or **NoSQL** is a group of information that does not follow a format, such as relational or tabular data. It can be presented as an image, video, metadata, transcripts, and so on. The data ingestion process usually involves a JSON file or a document collection, as we previously saw when ingesting data from **MongoDB**.

In this recipe, we will read a JSON file and transform it into a DataFrame without a schema. Although unstructured data is supposed to have a more flexible design, we will see some implications of not having any schema or structure in our DataFrame.

Getting ready...

Here, we will use the `holiday_brazil.json` file to create the DataFrame. You can find it in the GitHub repository here: `https://github.com/PacktPublishing/Data-Ingestion-with-Python-Cookbook`.

We will use `SparkSession` to read the JSON file and create a DataFrame to ensure the session is up and running.

All code can be executed in a Jupyter Notebook or at PySpark shell.

How to do it...

Let's now read our `holiday_brazil.json` file, observing how Spark handles it:

1. **Reading the JSON file**: Since our file has nested objects, we will pass `multiline` as a parameter to the `options()` method. We will also let PySpark infer the data types in it:

```
df_json = spark.read.option("multiline","true") \
                    .json('holiday_brazil.json')
```

If all goes right, you should see no output.

2. **Printing the inferred schema**: Using the `printSchema()` method, we can see how PySpark interpreted the data types of each key:

```
df_json.printSchema()
```

The following screenshot is the expected output:

```
root
 |-- holidays: array (nullable = true)
 |    |-- element: struct (containsNull = true)
 |    |    |-- country: string (nullable = true)
 |    |    |-- date: string (nullable = true)
 |    |    |-- name: string (nullable = true)
 |    |    |-- observed: string (nullable = true)
 |    |    |-- public: boolean (nullable = true)
 |    |    |-- uuid: string (nullable = true)
 |    |    |-- weekday: struct (nullable = true)
 |    |    |    |-- date: struct (nullable = true)
 |    |    |    |    |-- name: string (nullable = true)
 |    |    |    |    |-- numeric: string (nullable = true)
 |    |    |    |-- observed: struct (nullable = true)
 |    |    |    |    |-- name: string (nullable = true)
 |    |    |    |    |-- numeric: string (nullable = true)
 |-- requests: struct (nullable = true)
 |    |-- available: long (nullable = true)
 |    |-- resets: string (nullable = true)
 |    |-- used: long (nullable = true)
 |-- status: long (nullable = true)
 |-- warning: string (nullable = true)
```

Figure 6.10 – holiday_brazil.json DataFrame schema

3. **Visualizing it with Pandas**: Let's use the `toPandas()` function to visualize our DataFrame better:

```
df_json.toPandas()
```

You should see this output:

	holidays	requests	status	warning
0	[(BR, 2021-01-01, New Year's Day, 2021-01-01, ...	(9991, 2022-11-01 00:00:00, 9)	200	These results do not include state and provinc...

Figure 6.11 – Output of toPandas() vision from the DataFrame

How it works...

Let's take a look at the output from *step 2*:

```
root
 |-- holidays: array (nullable = true)
 |    |-- element: struct (containsNull = true)
 |    |    |-- country: string (nullable = true)
 |    |    |-- date: string (nullable = true)
 |    |    |-- name: string (nullable = true)
 |    |    |-- observed: string (nullable = true)
 |    |    |-- public: boolean (nullable = true)
 |    |    |-- uuid: string (nullable = true)
 |    |    |-- weekday: struct (nullable = true)
 |    |    |    |-- date: struct (nullable = true)
 |    |    |    |    |-- name: string (nullable = true)
 |    |    |    |    |-- numeric: string (nullable = true)
 |    |    |    |-- observed: struct (nullable = true)
 |    |    |    |    |-- name: string (nullable = true)
 |    |    |    |    |-- numeric: string (nullable = true)
 |-- requests: struct (nullable = true)
 |    |-- available: long (nullable = true)
 |    |-- resets: string (nullable = true)
 |    |-- used: long (nullable = true)
 |-- status: long (nullable = true)
 |-- warning: string (nullable = true)
```

Figure 6.12 – Holiday_brasil.json DataFrame schema

As we can observe, Spark only brought four columns, the first four keys in the JSON file, and ignored the rest of the other nested keys by keeping them inside the main four ones. This happens because Spark needs to handle the parameter better by flattening values from nested fields, even though we passed multiline as a parameter in the options() configuration.

Another important point is that data types inferred for the numeric keys inside the weekday array are string values and should be integers. This happened because those values have quotation marks, as you can see in the following figure:

```
"weekday":{
    "date":{
        "name":"Monday",
        "numeric":"1"
    },
    "observed":{
        "name":"Monday",
        "numeric":"1"
    }
}
```

Figure 6.13 – weekday objects

Applying this to a real-world scenario where it is crucial to have a qualitative evaluation, these unformatted and schemaless DataFrame can create problems later when uploaded to a data warehouse. If a field suddenly changes its name or is unavailable in the source, it can lead to data inconsistency. There are some ways to solve this, and we will cover this further in the following recipe, *Ingesting unstructured data with a well-defined schema and format*.

However, there are plenty of other scenarios where a schema is not needed to be defined or unstructured data doesn't need to be standardized. An excellent example is application logs or metadata, where data is often linked to the application's or system's availability to send the information. In that case, solutions such as **ElasticSearch**, **DynamoDB**, and many others are good storage options that provide query support. In other words, most of the issues here will be more inclined to generate a quantitative output.

Ingesting unstructured data with a well-defined schema and format

In the previous recipe, *Importing unstructured data without schema*, we read a JSON file without any schema or formatting application. This led us to an odd output, which could bring confusion and require additional work later in the data pipeline. While this example pertains specifically to a JSON file, it also applies to all other NoSQL or unstructured data that needs to be converted into analytical data.

The objective is to continue the last recipe and apply a schema and standard to our data, making it more legible and easy to process in the subsequent phases of **ETL**.

Getting ready

This recipe has the exact same requirements as the *Importing unstructured data without a schema* recipe.

How to do it...

We will perform the following steps to perform this recipe:

1. **Importing data types**: As usual, let's start by importing our data types from the PySpark library:

    ```
    from pyspark.sql.types import StructType, ArrayType,
    StructField, StringType, IntegerType, MapType
    ```

2. **Structuring the JSON schema**: Next, we set the schema based on how the JSON is structured:

    ```
    schema = StructType([ \
            StructField('status', StringType(), True),
            StructField('holidays', ArrayType(
                StructType([
                    StructField('name', StringType(), True),
                    StructField('date', DateType(), True),
                    StructField('observed', StringType(), True),
                    StructField('public', StringType(), True),
                    StructField('country', StringType(), True),
                    StructField('uuid', StringType(), True),
                    StructField('weekday', MapType(StringType(),
    MapType(StringType(),StringType(),True),True))
    ```

```
                    ])
            ))
      ])
```

3. **Applying the schema**: Similar to the CSV reading, we add the `schema()` method to apply the schema created in the previous step:

   ```
   df_json = spark.read.option("multiline","true") \
                    .schema(schema) \
                    .json('holiday_brazil.json')
   ```

4. **Expanding the columns**: Using the `explode()` method, let's broaden the fields inside the `holidays` column:

   ```
   from pyspark.sql.functions import explode
   exploded_json = df_json.select('status', explode("holidays").
   alias("holidaysExplode"))\
           .select("status", "holidaysExplode.*")
   ```

5. **Expanding more columns**: This is an optional step, but if needed, we can keep growing and flattening the other columns with nested content inside:

   ```
   exploded_json2 = exploded_json.select("*", explode('weekday').
   alias('weekday_type', 'weekday_objects'))
   ```

6. **Seeing our DataFrame**: Using the `toPandas()` function, we can better view what our DataFrame looks like now:

   ```
   exploded_json2.toPandas()
   ```

   You should see the following output:

	status	name	date	observed	public	country	uuid	weekday	weekday_type	weekday_objects
0	None	New Year's Day	2021-01-01	2021-01-01	true	BR	b58254f9-b38b-42c1-8b30-95a095798b0c	{'date': {'name': 'Friday', 'numeric': '5'}, ...	date	{'name': 'Friday', 'numeric': '5'}
1	None	New Year's Day	2021-01-01	2021-01-01	true	BR	b58254f9-b38b-42c1-8b30-95a095798b0c	{'date': {'name': 'Friday', 'numeric': '5'}, ...	observed	{'name': 'Friday', 'numeric': '5'}
2	None	Shrove Monday	2021-02-15	2021-02-15	false	BR	26346ac8-b1c7-4dfb-bda2-64c1ca445cc9	{'date': {'name': 'Monday', 'numeric': '1'}, ...	date	{'name': 'Monday', 'numeric': '1'}
3	None	Shrove Monday	2021-02-15	2021-02-15	false	BR	26346ac8-b1c7-4dfb-bda2-64c1ca445cc9	{'date': {'name': 'Monday', 'numeric': '1'}, ...	observed	{'name': 'Monday', 'numeric': '1'}
4	None	Shrove Tuesday	2021-02-16	2021-02-16	false	BR	4a764c30-0b8e-4182-b89d-edb419213c7b	{'date': {'name': 'Tuesday', 'numeric': '2'}, ...	date	{'name': 'Tuesday', 'numeric': '2'}
5	None	Shrove Tuesday	2021-02-16	2021-02-16	false	BR	4a764c30-0b8e-4182-b89d-edb419213c7b	{'date': {'name': 'Tuesday', 'numeric': '2'}, ...	observed	{'name': 'Tuesday', 'numeric': '2'}
6	None	Ash Wednesday	2021-02-17	2021-02-17	false	BR	1cd7828f-5fb6-4c18-a73d-e7dc444770bb	{'date': {'name': 'Wednesday', 'numeric': '3'}	date	{'name': 'Wednesday', 'numeric': '3'}

Figure 6.14 – The DataFrame with expanded columns

The final DataFrame can be saved as a Parquet file, which can make the next steps in the data pipeline easier to handle.

How it works...

As you can observe, this JSON file has some complexity due to the number of nested objects. When handling a file such as this, we can use several approaches. When coding, there are many approaches to reaching a solution. Let's understand how this recipe works.

In *step 1*, two new data types were imported—`ArrayType` and `MapType`. Even though there is some confusion when using each type, it is somewhat simple to understand when looking at the JSON structure.

We used `ArrayType` for the `holiday` key because its structure looks like this:

```
holiday : [{},{}...]
```

It is an array (or a list if we use Python) of objects, each representing a holiday in Brazil. When `ArrayType` has other objects inside it, we need to re-use `StructType` to inform Spark of the structure of objects that reside inside. That's why in *step 2*, our schema started to look like this:

```
StructField('holidays', ArrayType(
            StructType([
    ...
])
))
```

The following new data type is `MapType`. This type refers to an object with other objects inside. If we are only using Python, it can be referred to as a dictionary. PySpark extends this data type from a **superclass** in Spark, and you can read more about that here: https://spark.apache.org/docs/2.0.1/api/java/index.html?org/apache/spark/sql/types/MapType.html.

The syntax of `MapType` requires the key-value type, the value type, and whether it accepts null values. We used this type in the `weekday` field, as you can see here:

```
StructField('weekday', MapType(
StringType(),MapType(
StringType(),StringType(),True)
,True))
```

Maybe this is the most complex structure we have seen so far, and that was due to how JSON gets structured through it:

```
weekday : {
 day : {{}, {}},
 observed : {{}, {}}
}
```

The structure in our schema created `MapType` for `weekday` and the subsequent keys, `day` and `observed`.

Once our schema is defined and applied to our DataFrame, we can make a preview of it using `df_json.show()`. If the schema does not match the JSON structure, you should see this output:

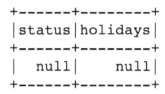

Figure 6.15 – df_json print

This demonstrates that Spark was unable to create the DataFrame correctly. The best action in this situation is to apply the schema step by step until the problem is resolved.

To flatten our fields inside the `holidays` column, we need to use a function from PySpark called `explode`, as you can see here:

```
from pyspark.sql.functions import explode
```

To apply the changes, we need to attribute the results to a variable where it will create a new DataFrame:

```
exploded_json = df_json.select('status', explode("holidays").
alias("holidaysExplode"))\
        .select("holidaysExplode.*")
```

> **Note**
>
> Although it seems redundant, preventing the original DataFrame from being modified is a good practice. If anything goes wrong, we don't need to reread the file because we have the intact state of the original DataFrame.

Using the `select()` method, we select the column that will remain intact and expand the desired one. We do so because Spark requires at least one column flattened as a reference. As you can observe, the other columns regarding the status of the **API** ingestion were removed from this schema.

The second parameter of `select()` is the `explode()` method, where we pass the `holiday` column and attribute an alias. The second chain, `select()`, will only retrieve the `holidaysExplode` column. *Step 5* follows the same process but for the `weekdays` column.

There's more...

As we previously discussed, there are many ways of flattening a JSON and applying a schema. You can see an example from Thomas at his *Medium blog* here: `https://medium.com/@thomaspt748/how-to-flatten-json-files-dynamically-using-apache-pyspark-c6b1b5fd4777`.

He uses Python functions to decouple the nested fields and then apply the PySpark code.

Towards Data Science also offers a solution using Python's lambda function. You can see it here: `https://towardsdatascience.com/flattening-json-records-using-pyspark-b83137669def`.

Flattening the JSON file can be a fantastic approach but requires more knowledge of complex Python functions. Again, there is no right way of doing it; it is important to bring a solution that you and your team can support.

See also

You can read more about the PySpark data structures here: `https://sparkbyexamples.com/pyspark/pyspark-structtype-and-structfield/`.

Inserting formatted SparkSession logs to facilitate your work

A commonly underestimated best practice is how to create valuable logs. Applications that log information and small code files can save a significant amount of debugging time. This is also true when ingesting or processing data.

This recipe approaches the best practice of logging events in our PySpark scripts. The examples here will give a more generic overview, which can be applied to any other piece of code and will even be used later in this book.

Getting ready

We will use the `listings.csv` file to execute the `read` method from Spark. You can find this dataset inside the GitHub repository for this book. Make sure your `SparkSession` is up and running.

How to do it...

Here are the steps to perform this recipe:

1. **Setting the log level**: Now, using `sparkContext`, we will assign the log level:

    ```
    spark.sparkContext.setLogLevel("INFO")
    ```

2. **Instantiating the log4j logger**: The next step is to create a variable that instantiates the
   getLogger() method:

```
Logger= spark._jvm.org.apache.log4j.Logger
syslogger = Logger.getLogger(__name__)
```

3. **Creating our logs**: With getLogger() instantiated, we can now call the internal methods
   representing the log levels, such as ERROR or INFO:

```
syslogger.error("Error message sample")
syslogger.info("Info message sample")
```

This gives us the following output:

```
23/01/04 11:11:53 ERROR __main__: Error message sample
23/01/04 11:11:53 INFO __main__: Info message sample
```

Figure 6.16 – Log message output example from Spark

4. **Creating a DataFrame**: Now, let's create a DataFrame using a file we have already seen in this
   chapter and observe the output:

```
try:
    df = spark.read.options(header=True, sep=',',
                             multiLine=True, escape='"',
                             inferSchema=True) \
                    .csv('listings.csv')
except Exception as e:
    syslogger.error(f"Error message: {e}")
```

You should see the following output:

```
23/01/04 11:29:20 INFO InMemoryFileIndex: It took 17 ms to list leaf files for 1 paths.
23/01/04 11:29:20 INFO MemoryStore: Block broadcast_0 stored as values in memory (estimated size 307.6 KiB, free 366.0 MiB)
23/01/04 11:29:20 INFO MemoryStore: Block broadcast_0_piece0 stored as bytes in memory (estimated size 27.6 KiB, free 366.0 Mi
B)
23/01/04 11:29:20 INFO BlockManagerInfo: Added broadcast_0_piece0 in memory on 172.31.148.225:46585 (size: 27.6 KiB, free: 366.
3 MiB)
23/01/04 11:29:20 INFO SparkContext: Created broadcast 0 from csv at NativeMethodAccessorImpl.java:0
23/01/04 11:29:20 INFO FileInputFormat: Total input files to process : 1
23/01/04 11:29:20 INFO FileInputFormat: Total input files to process : 1
23/01/04 11:29:20 INFO SparkContext: Starting job: csv at NativeMethodAccessorImpl.java:0
23/01/04 11:29:20 INFO DAGScheduler: Got job 0 (csv at NativeMethodAccessorImpl.java:0) with 1 output partitions
```

Figure 6.17 – PySpark logs output

Don't worry if more lines are showing in your console; the image was cut to make it more readable.

How it works...

Let's explore a bit of what was done in this recipe. Spark has a native library called `log4j`, or `rootLogger`. `log4j` is the default logging mechanism Spark uses to throw all log messages such as TRACE, DEBUG, INFO, WARN, ERROR, and FATAL. The severity of the message increases with each level. By default, Spark has logs at the WARN level, so new messages started to appear when we set it to INFO, such as memory store information.

We will observe more INFO logs when we execute a `spark-submit` command to run PySpark (we will cover this later in *Chapter 11*).

Depending on how big our script is and which environment it belongs to, it is a good practice to set it only to show ERROR messages. We can change the log level with the following code:

```
spark.sparkContext.setLogLevel("ERROR")
```

As you can observe, we used `sparkContext` to set the log level. `sparkContext` is a crucial component in Spark applications. It manages the cluster resources, coordinates the execution of tasks, and provides an interface for interacting with distributed datasets and performing computations in a distributed and parallel manner. Defining the log level of our code will prevent levels below ERROR from appearing on the console, making it cleaner:

```
Logger= spark._jvm.org.apache.log4j.Logger
syslogger = Logger.getLogger(__name__)
```

Next, we retrieved our `log4j` module and its `Logger` class from the instantiated session. This class has a method to show logs already formatted as Spark does. The `__name__` parameter will retrieve the name of the current module from the Python internals; in our case, it is `main`.

With this, we can create customized logs as follows:

```
syslogger.error("Error message sample")
syslogger.info("Info message sample")
```

And, of course, you can ally the PySpark logs with your Python output for a complete solution using a `try...except` exception-handling closure. Let's simulate an error by passing the wrong filename to our reading function as follows:

```
try:
    df = spark.read.options(header=True, sep=',',
                            multiLine=True, escape='"',
                            inferSchema=True) \
                 .csv('listings.cs') # Error here
except Exception as e:
    syslogger.error(f"Error message: {e}")
```

You should see the following output message:

```
23/01/04 11:59:11 ERROR __main__: Error message: Path does not exist: file:/home/glauesppen/listings.cs
```

Figure 6.18 – Error message formatted by log4j

There's more...

Many more customizations are available in log4j, but this might require a little more work. For example, you can change some WARN messages to appear as ERROR messages and prevent the script from continuing to be processed. A practical example in this chapter would be the *Importing structured data using a well-defined schema* recipe, when the number of columns in the schema does not match the file.

You can read more about it on *Ivan Trusov's* blog page here: https://polarpersonal.medium. com/writing-pyspark-logs-in-apache-spark-and-databricks-8590c28d1d51.

See also

- You can find more about the PySpark best practices here: https://climbtheladder. com/10-pyspark-logging-best-practices/.

- Read more about sparkContext and how it works in the Spark official documentation here: https://spark.apache.org/docs/3.2.0/api/java/org/apache/ spark/SparkContext.html.

Further reading

- https://www.tibco.com/reference-center/what-is-structured-data

- https://spark.apache.org/docs/latest/sql-programming-guide.html

- https://mungingdata.com/pyspark/schema-structtype-structfield/

- https://sparkbyexamples.com/pyspark/pyspark-select-nested-struct-columns/

- https://benalexkeen.com/using-pyspark-to-read-and-flatten-json-data-using-an-enforced-schema/

- https://towardsdatascience.com/json-in-databricks-and-pyspark-26437352f0e9

7
Ingesting Analytical Data

Analytical data is a bundle of data that serves various areas (such as finances, marketing, and sales) in a company, university, or any other institution, to facilitate decision-making, especially for strategic matters. When transposing analytical data to a data pipeline or a usual **Extract, Transform, and Load (ETL)** process, it corresponds to the final step, where data is already ingested, cleaned, aggregated, and has other transformations accordingly to business rules.

There are plenty of scenarios where data engineers must retrieve data from a data warehouse or any other storage containing analytical data. The objective of this chapter is to learn how to read analytical data and its standard formats and cover practical use cases related to the reverse ETL concept.

In this chapter, we will learn about the following topics:

- Ingesting Parquet files
- Ingesting Avro files
- Applying schemas to analytical data
- Filtering data and handling common issues
- Ingesting partitioned data
- Applying reverse ETL
- Selecting analytical data for reverse ETL

Technical requirements

Like *Chapter 6*, in this chapter too, some recipes will need `SparkSession` initialized, and you can use the same session for all of them. You can use the following code to create your session:

```
from pyspark.sql import SparkSession
spark = SparkSession.builder \
        .master("local[1]") \
        .appName("chapter7_analytical_data") \
```

```
.config("spark.executor.memory", '3g') \
.config("spark.executor.cores", '2') \
.config("spark.cores.max", '2') \
.getOrCreate()
```

> **Note**
>
> A `WARN` message as output is expected in some cases, especially if you are using WSL on Windows, so you don't need to worry if you receive one.

You can also find the code from this chapter in its GitHub repository here: `https://github.com/PacktPublishing/Data-Ingestion-with-Python-Cookbook`.

Ingesting Parquet files

Apache Parquet is a columnar storage format that is open source and designed to support fast processing. It is available to any project in a **Hadoop ecosystem** and can be read in different programming languages.

Due to its compression and fastness, this is one of the most used formats when needing to analyze data in great volume. The objective of this recipe is to understand how to read a collection of Parquet files using **PySpark** in a real-world scenario.

Getting ready

For this recipe, we will need `SparkSession` to be initialized. You can use the code provided at the beginning of this chapter to do so.

The dataset for this recipe will be *Yellow Taxi Trip Records from New York*. You can download it by accessing the **NYC Government website** and selecting **2022** | **January** | **Yellow Taxi Trip Records** or using this link:

`https://d37ci6vzurychx.cloudfront.net/trip-data/yellow_tripdata_2022-01.parquet`

Feel free to execute the code with a Jupyter notebook or a PySpark shell session.

How to do it...

Here are the steps to perform this recipe:

1. **Setting the Parquet file path**: To create a DataFrame based on the Parquet file, we have two options: pass the filename or the path of the Parquet file. In the following code block, you can see an example of passing only the name of the file:

   ```
   df = spark.read.parquet('chapter7_parquet_files/
   yellow_tripdata_2022-01.parquet')
   ```

   For the second option, we remove the Parquet filename, and Spark handles the rest. You can see how the code looks in the following code block:

   ```
   df = spark.read.parquet('chapter7_parquet_files/')
   ```

2. **Getting our DataFrame schema**: Now, let's use the `.printSchema()` function to see whether the DataFrame was created successfully:

   ```
   df.printSchema()
   ```

   You should see the following output:

   ```
   root
    |-- VendorID: long (nullable = true)
    |-- tpep_pickup_datetime: timestamp (nullable = true)
    |-- tpep_dropoff_datetime: timestamp (nullable = true)
    |-- passenger_count: double (nullable = true)
    |-- trip_distance: double (nullable = true)
    |-- RatecodeID: double (nullable = true)
    |-- store_and_fwd_flag: string (nullable = true)
    |-- PULocationID: long (nullable = true)
    |-- DOLocationID: long (nullable = true)
    |-- payment_type: long (nullable = true)
    |-- fare_amount: double (nullable = true)
    |-- extra: double (nullable = true)
    |-- mta_tax: double (nullable = true)
    |-- tip_amount: double (nullable = true)
    |-- tolls_amount: double (nullable = true)
    |-- improvement_surcharge: double (nullable = true)
    |-- total_amount: double (nullable = true)
    |-- congestion_surcharge: double (nullable = true)
    |-- airport_fee: double (nullable = true)
   ```

   Figure 7.1 – Yellow taxi trip DataFrame schema

3. **Visualizing with pandas**: This is an optional step since it requires your local machine to have enough processing capacity and can freeze your kernel trying to process it.

   To take a better look at the DataFrame, let's use `.toPandas()`, as shown:

   ```
   df.toPandas().head(10)
   ```

You should see the following output:

	VendorID	tpep_pickup_datetime	tpep_dropoff_datetime	passenger_count	trip_distance	RatecodeID	store_and_fwd_flag	PU
0	1	2022-01-01 00:35:40	2022-01-01 00:53:29	2.0	3.80	1.0	N	
1	1	2022-01-01 00:33:43	2022-01-01 00:42:07	1.0	2.10	1.0	N	
2	2	2022-01-01 00:53:21	2022-01-01 01:02:19	1.0	0.97	1.0	N	
3	2	2022-01-01 00:25:21	2022-01-01 00:35:23	1.0	1.09	1.0	N	
4	2	2022-01-01 00:36:48	2022-01-01 01:14:20	1.0	4.30	1.0	N	
5	1	2022-01-01 00:40:15	2022-01-01 01:09:48	1.0	10.30	1.0	N	
6	2	2022-01-01 00:20:50	2022-01-01 00:34:58	1.0	5.07	1.0	N	
7	2	2022-01-01 00:13:04	2022-01-01 00:22:45	1.0	2.02	1.0	N	
8	2	2022-01-01 00:30:02	2022-01-01 00:44:49	1.0	2.71	1.0	N	
9	2	2022-01-01 00:48:52	2022-01-01 00:53:28	1.0	0.78	1.0	N	

Figure 7.2 – Yellow taxi trip DataFrame with pandas visualization

How it works...

As we can observe in the preceding code, reading Parquet files is straightforward. Like many Hadoop tools, PySpark natively supports reading and writing Parquet.

Similar to JSON and CSV files, we used a function that derives from the `.read` method to inform PySpark that a Parquet file will be read, as follows:

```
spark.read.parquet()
```

We also saw two ways of reading, passing only the folder where the Parquet file is or passing the path with the Parquet filename. The best practice is to use the first case since there is usually more than one Parquet file, and reading just one may cause several errors. This is because each Parquet file inside the respective Parquet folder corresponds to a piece of the data.

After reading and transforming the dataset into a DataFrame, we printed its schema using the `.printSchema()` method. As the name suggests, it will print and show the schema of the DataFrame. Since we didn't specify the schema we want for the DataFrame, Spark will infer it based on the data pattern inside the columns. Don't worry about this now; we will cover this further in the *Applying schemas to analytical data* recipe.

Using the `.printSchema()` method before doing any operations in the DataFrame is an excellent practice for understanding the best ways to handle the data inside.

Finally, as the last step of the recipe, we used the `.toPandas()` method to visualize our data better since the `.show()` Spark method is not intended to bring friendly visualizations like pandas. However, we must be cautious when using the `.toPandas()` method since it needs computational and memory power to translate the Spark DataFrame into a pandas DataFrame.

There's more...

Now, let's understand why `parquet` is an optimized file format for big data. Parquet files are column-oriented, stored in data blocks and in small chunks (a data fragment), allowing optimized reading and writing. In the following diagram, you can visually observe how `parquet` organizes data at a high level:

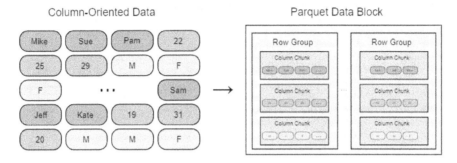

Figure 7.3 – Parquet file structure diagram by Darius Kharazi (https://dkharazi.github.io/blog/parquet)

Parquet files can frequently be found in a compressed form. This adds another layer of efficiency to improve data storage and transfer. A compressed Parquet file will look like this:

```
yellow_tripdata_2022-01.snappy.parquet
```

The `snappy` name informs us of the compression type and is crucial when creating a table in **AWS Athena**, **Impala**, or **Hive**. It is similar to the `gzip` format but more optimized for a massive volume of data.

See also

- You can read more about Apache Parquet in the official documentation: `https://parquet.apache.org/docs/overview/motivation/`

- If you want to test other Parquet files and explore more data from *Open Targets Platform*, access this link: `http://ftp.ebi.ac.uk/pub/databases/opentargets/platform/22.11/output/etl/parquet/hpo/`

Ingesting Avro files

Like Parquet, **Apache Avro** is a widely used format to store analytical data. Apache Avro is a leading method of serialization to record data and relies on schemas. It also provides **Remote Procedure Calls** (**RPCs**), making transmitting data easier and resolving problems such as missing fields, extra fields, and naming fields.

In this recipe, we will understand how to read an Avro file properly and later comprehend how it works.

Getting ready

This recipe will require `SparkSession` with some different configurations from the previous *Ingesting Parquet files* recipe. If you are already running `SparkSession`, stop it using the following command:

```
spark.stop()
```

We will create another session in the *How to do it...* section.

The dataset used here can be found at this link: `https://github.com/PacktPublishing/Data-Ingestion-with-Python-Cookbook/tree/main/Chapter_7/ingesting_avro_files`.

Feel free to execute the code in a Jupyter notebook or your PySpark shell session.

How to do it...

Here are the steps to perform this recipe:

1. **Creating a custom SparkSession**: To read an `avro` file, we need to import a `.jars` file in our `SparkSession` configuration, as follows:

    ```
    from pyspark.sql import SparkSession
    spark = SparkSession.builder \
            .master("local[1]") \
            .appName("chapter7_analytical_data_avro") \
            .config("spark.executor.memory", '3g') \
            .config("spark.executor.cores", '2') \
            .config("spark.cores.max", '2') \
            .config("spark.jars.packages", 'org.apache.spark:spark-
    avro_2.12:2.4.4') \
            .getOrCreate()
    ```

When executed, this code will provide an output similar to this:

```
Ivy Default Cache set to: /home/          /.ivy2/cache
The jars for the packages stored in: /home/          /.ivy2/jars
org.apache.spark#spark-avro_2.12 added as a dependency
:: resolving dependencies :: org.apache.spark#spark-submit-parent-8d3fb93d-3e80-4b6f-a5d2-76a3d630b546;1.0
        confs: [default]
        found org.apache.spark#spark-avro_2.12;2.4.4 in central
        found org.spark-project.spark#unused;1.0.0 in central
downloading https://repo1.maven.org/maven2/org/apache/spark/spark-avro_2.12/2.4.4/spark-avro_2.12-2.4.4.jar ...
        [SUCCESSFUL ] org.apache.spark#spark-avro_2.12;2.4.4!spark-avro_2.12.jar (21ms)
:: resolution report :: resolve 2671ms :: artifacts dl 22ms
        :: modules in use:
        org.apache.spark#spark-avro_2.12;2.4.4 from central in [default]
        org.spark-project.spark#unused;1.0.0 from central in [default]
        ---------------------------------------------------------------------
        |                  |            modules            ||   artifacts   |
        |       conf       | number| search|dwnlded|evicted|| number|dwnlded|
        ---------------------------------------------------------------------
        |     default      |   2   |   1   |   1   |   0   ||   2   |   1   |
        ---------------------------------------------------------------------
:: retrieving :: org.apache.spark#spark-submit-parent-8d3fb93d-3e80-4b6f-a5d2-76a3d630b546
        confs: [default]
        1 artifacts copied, 1 already retrieved (98kB/3ms)
23/01/18 20:44:37 WARN NativeCodeLoader: Unable to load native-hadoop library for your platform... using builtin-java classes w
here applicable
Using Spark's default log4j profile: org/apache/spark/log4j-defaults.properties
Setting default log level to "WARN".
To adjust logging level use sc.setLogLevel(newLevel). For SparkR, use setLogLevel(newLevel).
```

Figure 7.4 – SparkSession logs when downloading an Avro file package

It means the `avro` package was successfully downloaded and ready to use. We will later cover how it works.

2. **Creating the DataFrame**: With the `.jars` file configured, we will pass the file format to `.read` and add the file's path:

    ```
    df = spark.read.format('avro').load('chapter7_avro_files/file.
    avro')
    ```

3. **Printing the schema**: Using `.printSchema()`, let's retrieve the schema of this DataFrame:

    ```
    df.printSchema()
    ```

You will observe the following output:

```
root
 |-- vendorId: long (nullable = true)
 |-- tpep_pickup_datetime: timestamp (nullable = true)
 |-- tpep_dropoff_datetime: timestamp (nullable = true)
 |-- passenger_count: double (nullable = true)
 |-- trip_distance: double (nullable = true)
 |-- ratecodeId: double (nullable = true)
 |-- store_and_fwd_flag: string (nullable = true)
 |-- puLocationId: long (nullable = true)
 |-- doLocationId: long (nullable = true)
 |-- payment_type: long (nullable = true)
 |-- fare_amount: double (nullable = true)
 |-- extra: double (nullable = true)
 |-- mta_tax: double (nullable = true)
 |-- tip_amount: double (nullable = true)
 |-- tolls_amount: double (nullable = true)
 |-- improvement_surcharge: double (nullable = true)
 |-- total_amount: double (nullable = true)
 |-- congestion_surcharge: double (nullable = true)
 |-- airport_fee: double (nullable = true)
```

Figure 7.5 – DataFrame schema from the Avro file

As we can observe, this DataFrame contains the same data as the Parquet file covered in the last recipe, *Ingesting Parquet files*.

How it works...

As you can observe, we started this recipe slightly differently by creating `SparkSession` with a custom configuration. This is because, since version 2.4, Spark does not natively provide an internal API to read or write Avro files.

If you try to read the file used here without downloading the `.jars` file, you will get the following error message:

```
AnalysisException                         Traceback (most recent call last)
/tmp/ipykernel_554/1991426988.py in <module>
----> 1 df = spark.read.format('avro').load('chapter7_avro_files/')

~/.local/lib/python3.8/site-packages/pyspark/sql/readwriter.py in load(self, path, format, schema, **
options)
    202             self.options(**options)
    203             if isinstance(path, str):
--> 204                 return self._df(self._jreader.load(path))
    205             elif path is not None:
    206                 if type(path) != list:

~/.local/lib/python3.8/site-packages/py4j/java_gateway.py in __call__(self, *args)
   1302
   1303             answer = self.gateway_client.send_command(command)
-> 1304             return value = get_return_value(
   1305                 answer, self.gateway_client, self.target_id, self.name)
   1306

~/.local/lib/python3.8/site-packages/pyspark/sql/utils.py in deco(*a, **kw)
    115                 # Hide where the exception came from that shows a non-Pythonic
    116                 # JVM exception message.
--> 117                 raise converted from None
    118             else:
    119                 raise

AnalysisException: Failed to find data source: avro. Avro is built-in but external data source module
since Spark 2.4. Please deploy the application as per the deployment section of "Apache Avro Data Sou
rce Guide".
```

Figure 7.6 – Error message when Spark cannot find the Avro file package

Reading the error message, we can notice it is recommended to search for a third-party (or external) source called **Apache Avro Data Source**, so Spark will be able to install the required packages and read the avro file. Check out the Spark third-parties documentation, which can be found here: `https://spark.apache.org/docs/latest/sql-data-sources-avro.html#data-source-option`.

Even though the documentation has some helpful information about how to set it for different languages, such as **Scala** or Python, you might find the `org.apache.spark:spark-avro_2.12:3.3.1.jars` file incompatible with some PySpark versions, and so the recommendation is to use `org.apache.spark:spark-avro_2.12:2.4.4`.

Databricks, a company founded by the Spark creators, also has a public `.jars` file to be downloaded, but it is also incompatible with some versions of PySpark: `com.databricks:spark-avro_2.11:4.0.0`.

Due to the non-existence of an internal API to handle this file, we need to inform Spark of the format of the file, as shown here:

```
spark.read.format('avro')
```

We did the same thing when reading **MongoDB** collections, as seen in *Chapter 5*, in the *Ingesting data from MongoDB using PySpark* recipe.

Once our file is converted into a DataFrame, all other functionalities and operations are identical without prejudice. As we saw, Spark will infer the schema and transform it into a columnar format.

There's more...

Now that we know about both Apache Parquet and Apache Avro, you might wonder when to use each. Even though both are used to store analytical data, some key differences exist.

The main difference is how they store data. Parquet stores are in a columnar format, while Avro stores data in rows, which can be very efficient if you want to retrieve the entire row or dataset. However, columnar formats are much more optimized for aggregations or larger datasets, and `parquet` also supports more efficient queries using large-scale data and compression.

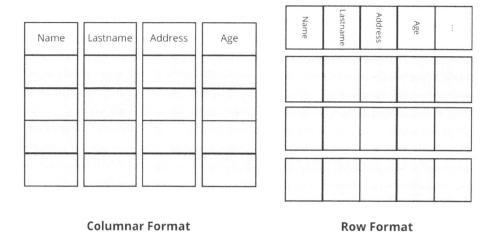

Columnar Format   **Row Format**

Figure 7.7 – Columnar versus row format

On the other hand, Avro files are commonly used for data streaming. A good example is when using **Kafka** with **Schema Registry**, it will allow Kafka to verify the file's expected schema in real time. You can see some example code in the Databricks documentation here: `https://docs.databricks.com/structured-streaming/avro-dataframe.html`.

See also

Read more about how Apache Avro works and its functionalities on the official documentation page here: `https://avro.apache.org/docs/`.

Applying schemas to analytical data

In the previous chapter, we saw how to apply schemas to structured and unstructured data, but the application of a schema is not limited to raw files.

Even when working with already processed data, there will be cases when we need to cast the values of a column or change column names to be used by another department. In this recipe, we will learn how to apply a schema to Parquet files and how it works.

Getting ready

We will need `SparkSession` for this recipe. Ensure you have a session that is up and running. We will use the same dataset as in the *Ingesting Parquet files* recipe.

Feel free to execute the code using a Jupyter notebook or your PySpark shell session.

How to do it...

Here are the steps to perform this recipe:

1. **Looking at our columns**: As seen in the *Ingesting Parquet files* recipe, we can list the columns and their inferred data types. You can see the list as follows:

    ```
    VendorID: long
    tpep_pickup_datetime: timestamp
    tpep_dropoff_datetime: timestamp
    passenger_count: double
    trip_distance: double
    RatecodeID: double
    store_and_fwd_flag: string
    PULocationID: long
    DOLocationID: long
    payment_type: long
    fare_amount: double
    extra: double
    mta_tax: double
    tip_amount: double
    tolls_amount: double
    improvement_surcharge: double
    total_amount: double
    congestion_surcharge: double
    ```

2. **Structuring our schema**: With the columns noted, it is easier to structure a schema using PySpark code. Let's use this opportunity to change the name of some columns, such as VendorID, to a more readable form. Refer to the following code:

```
from pyspark.sql.types import StructType, StructField,
StringType, LongType, DoubleType, TimestampType
schema = StructType([ \
    StructField("vendorId", LongType() ,True), \
    StructField("tpep_pickup_datetime", TimestampType() ,True),
\
    StructField("tpep_dropoff_datetime", TimestampType() ,True),
\
    StructField("passenger_count", DoubleType() ,True), \
    StructField("trip_distance", DoubleType() ,True), \
    StructField("ratecodeId", DoubleType() ,True), \
    StructField("store_and_fwd_flag", StringType() ,True), \
    StructField("puLocationId", LongType() ,True), \
    StructField("doLocationId", LongType() ,True), \
    StructField("payment_type", LongType() ,True), \
    StructField("fare_amount", DoubleType() ,True), \
    StructField("extra", DoubleType() ,True), \
    StructField("mta_tax", DoubleType() ,True), \
    StructField("tip_amount", DoubleType() ,True), \
    StructField("tolls_amount", DoubleType() ,True), \
    StructField("improvement_surcharge", DoubleType() ,True), \
    StructField("total_amount", DoubleType() ,True), \
    StructField("congestion_surcharge", DoubleType() ,True), \
    StructField("airport_fee", DoubleType() ,True), \
    ])
```

3. **Applying the schema**: With the schema stored in a variable, we can attribute it to a new DataFrame and reread the parquet file:

```
df_new_schema = spark.read.schema(schema).parquet('chapter7_
parquet_files/yellow_tripdata_2022-01.parquet')
```

4. **Printing the new DataFrame schema**: When printing the schema, we can see the name of some columns changed as we set them on the schema object in *step 1*:

```
df_new_schema.printSchema()
```

Executing the preceding code will provide the following output:

```
root
 |-- vendorId: long (nullable = true)
 |-- tpep_pickup_datetime: timestamp (nullable = true)
 |-- tpep_dropoff_datetime: timestamp (nullable = true)
 |-- passenger_count: double (nullable = true)
 |-- trip_distance: double (nullable = true)
 |-- ratecodeId: double (nullable = true)
 |-- store_and_fwd_flag: string (nullable = true)
 |-- puLocationId: long (nullable = true)
 |-- doLocationId: long (nullable = true)
 |-- payment_type: long (nullable = true)
 |-- fare_amount: double (nullable = true)
 |-- extra: double (nullable = true)
 |-- mta_tax: double (nullable = true)
 |-- tip_amount: double (nullable = true)
 |-- tolls_amount: double (nullable = true)
 |-- improvement_surcharge: double (nullable = true)
 |-- total_amount: double (nullable = true)
 |-- congestion_surcharge: double (nullable = true)
 |-- airport_fee: double (nullable = true)
```

Figure 7.8 – DataFrame with new schema applied

5. **Visualizing with pandas**: This step is optional since it may require some processing capacity from your local machine. We can double-check whether the final output is correct using the `.toPandas()` function, as follows:

```
df_new_schema.toPandas().head(10)
```

The output looks like this:

	vendorId	tpep_pickup_datetime	tpep_dropoff_datetime	passenger_count	trip_distance	ratecodeId	store_and_fwd_flag	puLc
0	1	2022-01-01 00:35:40	2022-01-01 00:53:29	2.0	3.80	1.0	N	
1	1	2022-01-01 00:33:43	2022-01-01 00:42:07	1.0	2.10	1.0	N	
2	2	2022-01-01 00:53:21	2022-01-01 01:02:19	1.0	0.97	1.0	N	
3	2	2022-01-01 00:25:21	2022-01-01 00:35:23	1.0	1.09	1.0	N	
4	2	2022-01-01 00:36:48	2022-01-01 01:14:20	1.0	4.30	1.0	N	
5	1	2022-01-01 00:40:15	2022-01-01 01:09:48	1.0	10.30	1.0	N	
6	2	2022-01-01 00:20:50	2022-01-01 00:34:58	1.0	5.07	1.0	N	
7	2	2022-01-01 00:13:04	2022-01-01 00:22:45	1.0	2.02	1.0	N	
8	2	2022-01-01 00:30:02	2022-01-01 00:44:49	1.0	2.71	1.0	N	
9	2	2022-01-01 00:48:52	2022-01-01 00:53:28	1.0	0.78	1.0	N	

Figure 7.9 – Yellow taxi trip DataFrame visualization with pandas

As you can see, no numerical data has changed, and therefore the data integrity remains.

How it works...

As we can observe in this exercise, defining and setting the schema for a DataFrame is not complex. However, it can be a bit laborious when we think about knowing the data types or declaring each column of the DataFrame.

The first step to start the schema definition is understanding the dataset we need to handle. This can be done by consulting a data catalog or even someone in more contact with the data. As a last option, you can create the DataFrame, let Spark infer the schema, and make adjustments when re-creating the DataFrame.

When creating the schema structure in Spark, there are a few items we need to pay attention to, as you can see here:

- **StructType**: The first thing to do is to declare the `StructType` object, which represents the schema of a list of `StructField`.

- **StructField**: `StructField` will define the name, data type, and whether the column allows null or empty fields.

- **Data types**: The last thing to bear in mind is where we will define the column's data type; as you can imagine, a few data types are available. You can always consult the documentation to see the supported data types here: `https://spark.apache.org/docs/latest/sql-ref-datatypes.html`.

Once we have defined the schema object, we can easily attribute it to the function that creates the DataFrame using the `.schema()` method, as we saw in *step 3*.

With the DataFrame created, all the following commands remain the same.

There's more...

Let's do an experiment where instead of using `TimestampType()`, we will use `DateType()`. See the following portion of the changed code:

```
StructField("tpep_pickup_datetime", DateType() ,True), \
StructField("tpep_dropoff_datetime", DateType() ,True), \
```

If we repeat the steps using the preceding code change, an error message will appear when we try to visualize the data:

```
23/05/01 11:22:50 ERROR Executor: Exception in task 0.0 in stage 0.0 (TID 0)
java.lang.NullPointerException
        at org.apache.spark.sql.execution.vectorized.OnHeapColumnVector.putLong(OnHeapColumnVector.j
ava:327)
        at org.apache.spark.sql.execution.datasources.parquet.VectorizedColumnReader.decodeDictionar
yIds(VectorizedColumnReader.java:418)
        at org.apache.spark.sql.execution.datasources.parquet.VectorizedColumnReader.readBatch(Vecto
rizedColumnReader.java:280)
        at org.apache.spark.sql.execution.datasources.parquet.VectorizedParquetRecordReader.nextBatc
h(VectorizedParquetRecordReader.java:283)
        at org.apache.spark.sql.execution.datasources.parquet.VectorizedParquetRecordReader.nextKeyV
alue(VectorizedParquetRecordReader.java:181)
        at org.apache.spark.sql.execution.datasources.RecordReaderIterator.hasNext(RecordReaderItera
tor.scala:37)
        at org.apache.spark.sql.execution.datasources.FileScanRDD$$anon$1.hasNext(FileScanRDD.scala:
93)
        at org.apache.spark.sql.execution.datasources.FileScanRDD$$anon$1.nextIterator(FileScanRDD.s
cala:173)
        at org.apache.spark.sql.execution.datasources.FileScanRDD$$anon$1.hasNext(FileScanRDD.scala:
```

Figure 7.10 – Error reading the DataFrame when attributing an incompatible data type

The reason behind this is the incompatibility of these two data types when formatting the data inside the column. `DateType()` uses the `yyyy-MM-dd` format, while `TimestampType()` uses `yyy-MM-dd HH:mm:ss.SSSS`.

Looking closely at both columns, we see hour, minute, and second information. If we try to force it into another format, it could corrupt the data.

See also

Learn more about the Spark data types here: `https://spark.apache.org/docs/3.0.0-preview/sql-ref-datatypes.html`.

Filtering data and handling common issues

Filtering data is a process of excluding or selecting only the necessary information to be used or stored. Even analytical data must be re-filtered to meet a specific need. An excellent example is **data marts** (we will cover them later in this recipe).

This recipe aims to understand how to create and apply filters to our data using a real-world example.

Getting ready

This recipe requires `SparkSession`, so ensure yours is up and running. You can use the code provided at the beginning of the chapter or create your own.

The dataset used here will be the same as in the *Ingesting Parquet files* recipe.

To make this exercise more practical, let's imagine we want to analyze two scenarios: how many trips each vendor made and what hour of the day there are more pickups. We will create some aggregations and filter our dataset to carry out those analyses.

How to do it...

Here are the steps to perform this recipe:

1. **Reading the Parquet file**: Let's start by reading our Parquet file, as the following code shows:

```
df = spark.read.parquet('chapter7_parquet_files/')
```

2. **Getting the vendor in ascending order**: Let's imagine a scenario where we want to filter all the vendorId instances and how many trips each vendor carried out in January (the timeframe of our dataset). We can use a .groupBy() function with .count(), as follows:

```
df.groupBy("vendorId").count().orderBy("vendorId").show()
```

This is what the vendor trip count looks like:

```
+--------+-------+
|vendorId|  count|
+--------+-------+
|       1| 742273|
|       2|1716059|
|       5|     36|
|       6|   5563|
+--------+-------+
```

Figure 7.11 – vendorId trips count

3. **Retrieving the busiest pickup hours**: Now, let's group by the hours of the pickups using the tpep_pickup_datetime column, as shown in the following code. Then, we make a count and order it in an ascending flow:

```
from pyspark.sql.functions import year, month, dayofmonth
df.groupBy(hour("tpep_pickup_datetime")).count().
orderBy("count").show()
```

This is what the output looks like:

```
+--------------------------+------+
|hour(tpep_pickup_datetime)| count|
+--------------------------+------+
|                         4| 12827|
|                         5| 14444|
|                         3| 19308|
|                         2| 29193|
|                         6| 36206|
|                         1| 42313|
|                         0| 60065|
|                         7| 74137|
|                        23| 80209|
|                         8|101528|
|                        22|104649|
|                         9|109376|
|                        21|109629|
|                        20|117995|
|                        10|119816|
|                        11|129561|
|                        12|142216|
|                        13|147878|
|                        19|151346|
|                        14|163858|
+--------------------------+------+
only showing top 20 rows
```

Figure 7.12 – Count of trips per hour

As you can observe, at 14 hours and 19 hours, there is an increase in the number of pickups. We can think of some possible reasons for this, such as lunchtime and rush hour.

How it works...

Since we are already very familiar with the reading operation to create a DataFrame, let's go straight to *step 2*:

```
df.groupBy("vendorId").count().orderBy("vendorId").show()
```

As you can observe, the chain of functions here closely resembles SQL operations. This is because, behind the scenes, we are using the native SQL methods a DataFrame supports.

> **Note**
> Like the SQL operations, the order of the methods in the chain will influence the result, and even whether it will result in a success or not.

In *step 3*, we added a little bit more complexity by extracting the hour value from the `tpep_pickup_datetime` column. That was only possible because this column is of the timestamp data type. Also, we ordered by the count column this time, which was created once we invoked the `.count()` function, similar to the SQL. You can see this in the following code:

```
df.groupBy(hour("tpep_pickup_datetime")).count().orderBy("count").
show()
```

There are plenty of other functions, such as `.filter()` and `.select()`. You can find more PySpark functions here: `https://spark.apache.org/docs/latest/api/python/reference/pyspark.sql/functions.html`.

There's more...

The analysis in this recipe was carried out using SQL functions natively supported by PySpark. However, these functions are not a good fit for more complex queries.

In those cases, the best practice is to use the **SQL API** of Spark. Let's see how to do it in the code that follows:

1. **Creating a temporary view**: First, we will create a temporary view based on our DataFrame. For this, we use the `.createOrReplaceTempView()` method and pass a name to our temporary view, as follows:

    ```
    df.createOrReplaceTempView("ny_yellow_taxi_data")
    ```

2. **Inserting our SQL query**: Using the `SparkSession` variable (`spark`), we will invoke `.sql()` and pass a multi-lined string containing the desired SQL code. To make it easier to visualize the results, let's also attribute it to a variable called `vendor_groupby`.

    Observe we use the temporary view name to indicate where the query will be made:

    ```
    vendor_groupby = spark.sql(
    """
    SELECT vendorId, COUNT(*) FROM ny_yellow_taxi_data
    GROUP BY vendorId
    ORDER BY COUNT(*)
    """
    )
    ```

    Executing this code will not generate an output.

3. **Showing our results**: Once the query is executed and instantiated on a variable, it will be a Spark DataFrame object, and the `.show()` method will work to bring the results, as shown in the following code:

    ```
    vendor_groupby.show()
    ```

This is the output:

```
+--------+--------+
|vendorId|count(1)|
+--------+--------+
|       5|      36|
|       6|    5563|
|       1|  742273|
|       2| 1716059|
+--------+--------+
```

Figure 7.13 – vendorId counts of trips using SQL code

The downside of using the SQL API is that the error logs might sometimes be unclear. See the following screenshot:

```
AnalysisException: expression 'ny_yellow_taxi_data.`tpep_pickup_datetime`' is neither present in the
group by, nor is it an aggregate function. Add to group by or wrap in first() (or first_value) if you
don't care which value you get.;
Aggregate [vendorId#0L], [VendorID#0L, tpep_pickup_datetime#1, tpep_dropoff_datetime#2, passenger_cou
nt#3, trip_distance#4, RatecodeID#5, store_and_fwd_flag#6, PULocationID#7L, DOLocationID#8L, payment_
type#9L, fare_amount#10, extra#11, mta_tax#12, tip_amount#13, tolls_amount#14, improvement_surcharge#
15, total_amount#16, congestion_surcharge#17, airport_fee#18]
+- SubqueryAlias ny_yellow_taxi_data
   +- Relation[VendorID#0L,tpep_pickup_datetime#1,tpep_dropoff_datetime#2,passenger_count#3,trip_dist
ance#4,RatecodeID#5,store_and_fwd_flag#6,PULocationID#7L,DOLocationID#8L,payment_type#9L,fare_amount#
10,extra#11,mta_tax#12,tip_amount#13,tolls_amount#14,improvement_surcharge#15,total_amount#16,congest
ion_surcharge#17,airport_fee#18] parquet
```

Figure 7.14 – Spark error when SQL does not have the right syntax

This screenshot shows the result of a query where the syntax was incorrect when grouping by the `tpep_pickup_datetime` column. In scenarios like this, the best approach is to debug using baby steps, executing the query operations and conditionals one by one. If your DataFrame comes from a table in a database, try to reproduce the query directly on the database and see whether there is a more intuitive error message.

Data marts

As mentioned at the beginning of this recipe, one common use case for ingesting and re-filtering analytical data is to use it in a data mart.

Data marts are a smaller version of a data warehouse, with data concentrated on one subject, such as from a financial or sales department. The following diagram shows how they tend to be organized:

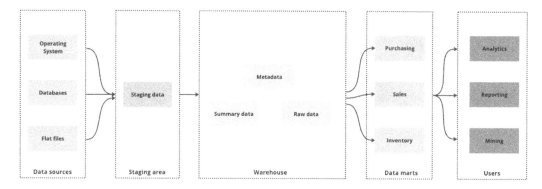

Figure 7.15 – Data marts diagram

Implementing the data mart concept has many benefits, such as reaching specific information or guaranteeing temporary access to a strict piece of data for a project without managing the security access of several users to the data warehouse.

See also

Find out more about data marts and data warehouse concepts on the Panoply.io blog: `https://panoply.io/data-warehouse-guide/data-mart-vs-data-warehouse/`.

Ingesting partitioned data

The practice of partitioning data is not recent. It was implemented in databases to distribute data across multiple disks or tables. Actually, data warehouses can partition data according to the purpose and use of the data inside. You can read more here: `https://www.tutorialspoint.com/dwh/dwh_partitioning_strategy.htm`.

In our case, partitioning data is related to how our data will be split into small chunks and processed.

In this recipe, we will learn how to ingest data that is already partitioned and how it can affect the performance of our code.

Getting ready

This recipe requires an initialized `SparkSession`. You can create your own or use the code provided at the beginning of this chapter.

The data required to complete the steps can be found here: `https://github.com/PacktPublishing/Data-Ingestion-with-Python-Cookbook/tree/main/Chapter_7/ingesting_partitioned_data`.

You can use a Jupyter notebook or a PySpark shell session to execute the code.

How to do it...

Use the following steps to perform this recipe:

1. **Creating the DataFrame for the February data**: Let's use the usual way for creating a DataFrame from a Parquet file, but this time passing only the month we want to read:

```
df_partitioned = spark.read.parquet("chapter7_partitioned_files/
month=2/")
```

You should see no output from this execution.

2. **Using pandas to show the results**: Once the DataFrame is created, we can better visualize the results using pandas:

```
df_partitioned.toPandas()
```

You should observe this output:

	VendorID	tpep_pickup_datetime	tpep_dropoff_datetime	passenger_count	trip_distance	RatecodeID	store_and_fwd_flag	PUI
0	2	2022-02-01 00:00:23	2022-02-01 00:17:08	1.0	7.98	1.0	N	
1	2	2022-02-01 03:00:05	2022-02-01 03:15:08	3.0	2.62	1.0	N	
2	2	2022-02-01 00:01:20	2022-02-01 00:24:59	1.0	10.83	1.0	N	
3	2	2022-02-01 03:30:04	2022-02-01 03:34:05	3.0	0.85	1.0	N	
4	2	2022-02-01 00:00:46	2022-02-01 00:15:43	5.0	3.71	1.0	N	
5	2	2022-02-22 11:31:12	2022-02-22 11:42:30	3.0	1.50	1.0	N	
6	2	2022-02-01 00:01:08	2022-02-01 00:20:08	5.0	7.33	1.0	N	

Figure 7.16 – Yellow taxi trip DataFrame visualization using partitioned data

Observe that in the `tpep_pickup_datetime` column, there is only data from February, and now we don't need to be very preoccupied with the processing capacity of our local machine.

How it works...

This was a very simple recipe, but there are some important concepts that we need to understand a bit.

As you can observe, all the magic happens during the creation of the DataFrame, where we pass not only the path where our Parquet files are stored but also the name of another subfolder containing the month reference. Let's take a look at how this folder is organized in the following screenshot:

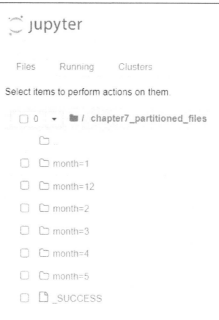

Figure 7.17 – Folder showing data partitioned by month

> **Note**
>
> The _SUCCESS file indicates that the partitioning write process was successfully made.

Inside the `chapter7_partitioned_files` folder, there are other subfolders with a number of references. Each of these subfolders represents a partition, in this case, categorized by month.

If we look inside a subfolder, we can observe one or more Parquet files. Refer to the following screenshot:

Figure 7.18 – Parquet file for February

Partitions are an optimized form of reading or writing a specific amount of data from a dataset. That's why using pandas to visualize the DataFrame was faster this time.

Partitioning also makes the execution of transformations faster since data will be processed using parallelism across the Spark internal worker nodes. You can visualize it better in the following figure:

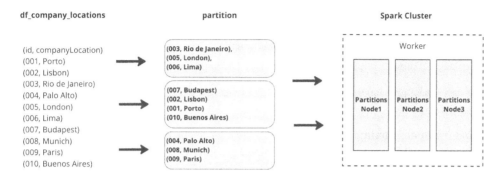

Figure 7.19 – Partitioning parallelism diagram

There's more...

As we saw, working with partitions to save data on a large scale brings several benefits. However, knowing how to partition your data is the key to reading and writing data in a performative way. Let's list the three most important best practices when writing partitions:

- **Use the right columns to partition**: In this exercise, we saw an example of partitioning by the column called month, but it is possible to partition over any column and even to use a column with year, month, or day to bring more granularity. Normally, partitioning reflects what the best way to retrieve the data will be.

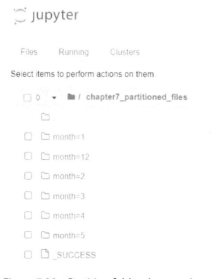

Figure 7.20 – Partition folders by month

- **Coalesce small files**: Depending on how our `SparkSession` is configured or where it will write the final files, Spark can create small `parquet/avro` files. From a large data scale perspective, reading these small files can prejudice performance.

  A good practice is to use `coalesce()` while invoking the `write()` function to aggregate the files into a small amount. Here is an example:

  ```
  df.coalesce(1).write.parquet('myfile/path')
  ```

  You can find a good article about it here: `https://www.educba.com/pyspark-coalesce/`.

- **Avoid over-partitioning**: This follows the same logic as the previous one. Over-partitioning will create small files since we split them using a granularity rule, and then Spark's parallelism will be slowed.

See also

You can find more good practices here: `https://climbtheladder.com/10-spark-partitioning-best-practices/`.

Related to the topic of partitioning, we also have the *database sharding* concept. It is a very interesting topic, and the MongoDB official documentation has a very nice post about it here: `https://www.mongodb.com/features/database-sharding-explained`.

Applying reverse ETL

As the name suggests, **reverse ETL** takes data from a data warehouse and inserts it into a business application such as **HubSpot** or **Salesforce**. The reason behind this is to make data more operational and use business tools to bring more insights to data that is already in a format ready for analysis or analytical format.

This recipe will teach us how to architect a reverse ETL pipeline and about the commonly used tools.

Getting ready

There are no technical requirements for this recipe. However, it is encouraged to use a whiteboard or a notepad to take notes.

Here, we will work with a scenario where we are ingesting data from an **e-learning platform**. Imagine we received a request from the marketing department to better understand user actions on the platform using the Salesforce system.

The objective here will be to create a diagram showing the data flow process from a source of data to the Salesforce platform.

How to do it...

To make this exercise more straightforward, we will assume we already have data stored in the database for the e-learning platform:

1. **Ingesting user action data from the website**: Let's imagine we have a frontend API that sends useful information about our user's actions and behavior on the e-learning platform to our backend databases. Refer to the following diagram to see what it looks like:

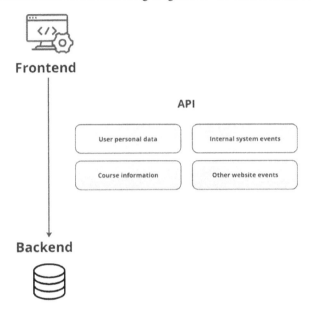

Figure 7.21 – Data flow from the frontend to an API in the backend

2. **Processing it using ETL**: With the available data, we can pick the necessary information that the marketing department needs and put it into the ETL process. We will ingest it from the backend database, apply any cleansing or transformations needed, and then load it into our data warehouse.

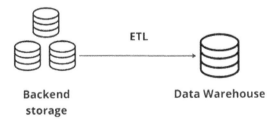

Figure 7.22 – Diagram showing backend storage to the data warehouse

3. **Storing data in a data warehouse**: After the data is ready and transformed into an analytical format, it will be stored in the data warehouse. We don't need to worry here about how data is modeled. Let's assume a new analytical table will be created just for this processing purpose.

4. **ETL to Salesforce**: Once data is populated in the data warehouse, we need to insert it into the Salesforce system. Let's do this using PySpark, as you can see in the following figure:

Figure 7.23 – Data warehouse data flow to Salesforce

With data inside Salesforce, we can advise the marketing team and automate this process to be triggered on a necessary schedule.

How it works...

Although it seems complicated, the reverse ETL process is similar to an ingest job. In some cases, adding a few more transformations to fit the final application model might be necessary, but isn't complex. Now, let's take a closer look at what we did in the recipe.

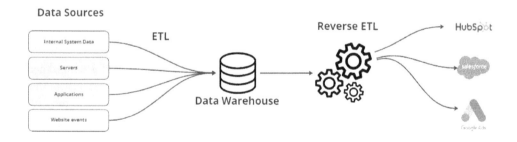

Figure 7.24 – Reverse ETL diagram overview

First, we need to understand whether we already have the requested raw data stored in our internal database to meet the marketing department's needs. If we don't, the data team is responsible for reaching out to the responsible developers to verify how to accomplish that.

Once this is checked, we proceed with the usual ETL pipeline. Normally, there will be SQL transformations to filter or group information based on the needs of the analysis. Then, we store it in a *source of truth*, such as a main data warehouse.

It is in *step 4* that the reverse ETL occurs. The origin of this name is because, normally, the ETL process involves retrieving data from an application such as Salesforce and storing it in a data warehouse. However, in recent years, these tools have become another form of better understanding how users are behaving or interacting with our applications.

With user-centric feedback solutions with analytical data, we can get better insights into and access to specific results. Another example besides Salesforce can be a **machine learning** solution to predict whether some change in the e-learning platform would result in improved user retention.

There's more...

To carry out reverse ETL, we can create our own solution or use a commercial one. Plenty of solutions on the market retrieve data from data warehouses and connect dynamically with business solutions. Some can also generate reports to provide feedback to the data warehouse again, improving the quality of information sent and even creating other analyses. The cons of these tools are that most are paid solutions, and free tiers tend to include one or few connections.

One of the most used reverse ETL tools is **Hightouch**; you can find out more here: `https://hightouch.com/`.

See also

You can read more about reverse ETL at *Astasia Myers'* Medium blog: `https://medium.com/memory-leak/reverse-etl-a-primer-4e6694dcc7fb`.

Selecting analytical data for reverse ETL

Now that we know what reverse ETL is, the next step is to understand which types of analytical data are a good use case to load into a Salesforce application, for example.

This recipe continues from the previous one, *Applying reverse ETL*, intending to illustrate a real scenario of deciding what data will be transferred into a Salesforce application.

Getting ready

This recipe has no technical requirements, but you can use a whiteboard or a notepad for annotations.

Still using the example of a scenario where the marketing department requested data to be loaded into their Salesforce account, we will now go a little deeper to see what information is relevant for their analysis.

We received a request from the marketing team to understand the user journey in the e-learning platform. They want to understand which courses are watched most and whether some need improvement. Currently, they don't know what information we have in our backend databases.

How to do it...

Let's work on this scenario in small steps. The objective here will be to understand what data we need to accomplish the task:

1. **Consulting the data catalog**: To simplify our work, let's assume our data engineers worked on creating a data catalog with the user information collected and stored in the backend databases. In the following diagram, we can better see how the information is stored:

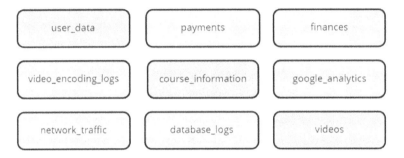

Figure 7.25 – Tables of interest for reverse ETL highlighted

As we can see, there are three tables with potentially relevant information to be retrieved: user_data, course_information, and videos.

2. **Selecting the raw data**: We can see highlighted in the following diagram the columns that can provide the information needed for the analysis:

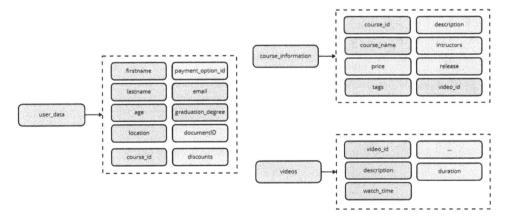

Figure 7.26 – Tables and respective columns highlighted as relevant for reverse ETL

3. **Transforming and filtering data**: Since we need a single table to load data into Salesforce, we can make a SQL filter and join the information.

How it works...

As mentioned at the beginning of the recipe, the marketing team wants to understand the user journey in the e-learning application. First, let's understand what a user journey is.

A user journey is all the actions and interactions a user carries out on a system or application, from when they opt to use or buy a service until they log out or leave it. Information such as what type of content they have watched and for how long is very important in this case.

Let's see the fields we collected and why they are important. The first six columns will give us an idea of the user and where they live. We can use these pieces of information later to see whether there are any patterns for the predilection of content.

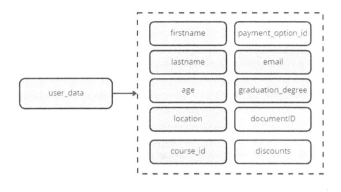

Figure 7.27 – user_data relevant columns for reverse ETL

Then, the last columns provide information about the content this user is watching and whether there is any relationship between the types of content. For example, if they bought a Python course and a SQL course, we can use a tag (for example, `programming course`) from the content metadata to make a filer of correlation.

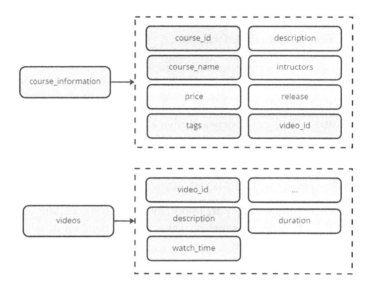

Figure 7.28 – course_information and videos with columns highlighted

Feeding back all this information into Salesforce can help to answer the following questions about the user's journey:

- Is there a tendency to finish one course before starting another?

- Do users tend to watch multiple courses at the same time?

- Does the educational team need to reformulate a course because of a high turnover?

See also

Find more use cases here: https://www.datachannel.co/blogs/reverse-etl-use-cases-common-usage-patterns.

Further reading

Here is a list of websites you can refer to, to enhance your knowledge further:

- https://segment.com/blog/reverse-etl/

- https://hightouch.com/blog/reverse-etl

- https://www.oracle.com/br/autonomous-database/what-is-data-mart/

- https://www.netsuite.com/portal/resource/articles/ecommerce/customer-lifetime-value-clv.shtml

- https://www.datachannel.co/blogs/reverse-etl-use-cases-common-usage-patterns

Part 2: Structuring the Ingestion Pipeline

In the book's second part, you will be introduced to the monitoring practices and see how to create your very first data ingestion pipeline using the recommended tools on the market, all the while applying the best data engineering practices.

This part has the following chapters:

- *Chapter 8, Designing Monitored Data Workflows*
- *Chapter 9, Putting Everything Together with Airflow*
- *Chapter 10, Logging and Monitoring Your Data Ingest in Airflow*
- *Chapter 11, Automating Your Data Ingestion Pipelines*
- *Chapter 12, Using Data Observability for Debugging, Error Handling, and Preventing Downtime*

8

Designing Monitored Data Workflows

Logging code is a good practice that allows developers to debug faster and provide maintenance more effectively for applications or systems. There is no strict rule when inserting logs, but knowing when not to spam your monitoring or alerting tool while using it is excellent. Creating several logging messages unnecessarily will obfuscate the instance when something significant happens. That's why it is crucial to understand the best practices when inserting logs into code.

This chapter will show how to create efficient and well-formatted logs using Python and PySpark for data pipelines with practical examples that can be applied in real-world projects.

In this chapter, we have the following recipes:

- Inserting logs
- Using log-level types
- Creating standardized logs
- Monitoring our data ingest file size
- Logging based on data
- Retrieving SparkSession metrics

Technical requirements

You can find the code from this chapter in the GitHub repository at `https://github.com/PacktPublishing/Data-Ingestion-with-Python-Cookbook`.

Inserting logs

As mentioned in the introduction of this chapter, adding logging functionality to your applications is essential for debugging or making improvements later on. However, creating several log messages without necessity may generate confusion or even cause us to miss crucial alerts. In any case, knowing which kind of message to show is indispensable.

This recipe will cover how to create helpful log messages using Python and when to insert them.

Getting ready

We will use only Python code. Make sure you have Python version 3.7 or above. You can use the following command to check it on your **command-line interface (CLI)**:

```
$ python3 --version
Python 3.8.10
```

The following code execution can be done on a Python shell or a Jupyter notebook.

How to do it...

To perform this exercise, we will make a function that reads and returns the first line of a CSV file using the best logging practices. Here is how we do it:

1. First, let's import the libraries we will use and set the primary configuration for our `logging` library:

    ```
    import csv
    import logging

    logging.basicConfig(filename='our_application.log',
    level=logging.INFO)
    ```

    Notice that we passed a filename parameter to the `basicConfig` method. Our logs will be stored there.

2. Next, we will create a simple function to read and return a CSV file's first line. Observe that `logging.info()` calls are inserted inside the functions with a message, as follows:

    ```
    def get_csv_first_line (csv_file):
        logging.info(f"Starting function to read first line")
        try:
            with open(csv_file, 'r') as file:
                logging.info(f" Opening and reading the CSV file")
                reader = csv.reader(file)
                first_row = next(reader)
    ```

```
            return first_row
    except Exception as e:
        logging.error(f"Error when reading the CSV file: {e}")
        raise
```

3.   Then, let's call our function, passing a CSV file as an example. Here, I will use the `listings.csv` file, which you can find in the GitHub repository as follows:

```
get_csv_first_line("listings.csv")
```

You should see the following output:

```
['id',
 'name',
 'host_id',
 'host_name',
 'neighbourhood_group',
 'neighbourhood',
 'latitude',
 'longitude',
 'room_type',
 'price',
 'minimum_nights',
 'number_of_reviews',
 'last_review',
 'reviews_per_month',
 'calculated_host_listings_count',
 'availability_365',
 'number_of_reviews_ltm',
 'license']
```

Figure 8.1 – gets_csv_first_line function output

4.   Let's check the directory where we executed our Python script or Jupyter notebook. You should see a file named `our_application.log`. If you click on it, the result should be as follows:

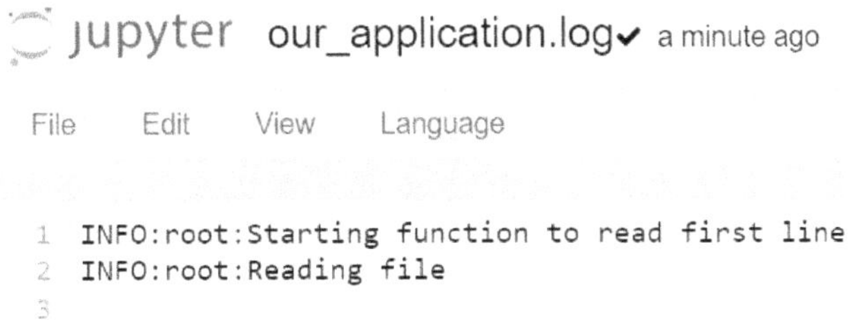

Figure 8.2 – our_application.log content

As you can see, we had two different outputs: one with the function results (*step 3*) and another that creates a file containing the log messages (*step 4*).

How it works...

Let's start understanding how the code works by looking at the first lines:

```
import logging
logging.basicConfig(filename='our_application.log', level=logging.
INFO)
```

After importing the built-in logging library, we called a method named `basicConfig()`, which sets the primary configuration for the subsequent calls in our function. The `filename` parameter indicates we want to save the logs into a file, and the `level` parameter sets the log level at which we want to start seeing messages. This will be covered in more detail in the *Using log-level types* recipe later in this chapter.

Then, we proceeded by creating our function and inserting the logging calls. Looking closely, we inserted the following:

```
logging.info(f"Starting function to read first line")
logging.info(f"Opening and reading the CSV file")
```

These two logs are informative and track an action or inform us as we pass through a part of the code or module. The best practice is to keep it as clean and objective as possible, so the next person (or even yourself) can identify where to start to look to solve a problem.

The next log informs us of an error, as you can see here:

```
logging.error(f"Error when reading the CSV file: {e}")
```

The way to make the call for this error method is similar to the `.info()` ones. In this case, the best practice is to use only exception clauses and pass the error as a string function, as we did by passing the e variable in curly brackets. This way, even if we cannot see the Python traceback, we will store it in a file or monitoring application.

> **Note**
>
> It is a common practice to encapsulate the exception output in a variable, as in `except Exception as e`. It allows us to control how we show or get a part of the error message.

Since our function was executed successfully, we don't expect to see any error message in our `our_application.log` file, as you can see here:

```
INFO:root:Starting function to read first line
INFO:root:Reading file
```

If we look closely at the structure of the saved log, we will notice a pattern. The first word on each line, `INFO`, indicates the log level; after this, we see the `root` word, which indicates the logging hierarchy; and finally, we get the message we inserted into the code.

We can optimize and format our logs in many ways, but we won't worry about this for now. We will cover the logging hierarchy in more detail in the *Formatting logs* recipe.

See also

See more about initiating the logs in Python in the official documentation here: `https://docs.python.org/3/howto/logging.html#logging-to-a-file`

Using log-level types

Now that we have been introduced to how and where to insert logs, let's understand log types or levels. Each log level has its own degree of relevance inside any system. For instance, the console output does not show debug messages by default.

We already covered how to log levels using PySpark in the *Inserting formatted SparkSession logs to facilitate your work recipe* in *Chapter 6*. Now we will do the same using only Python. This recipe aims to show how to set logging levels at the beginning of your script and insert the different levels inside your code to create a hierarchy of priority for your logs. With this, you can create a structured script that allows you or your team to monitor and identify errors.

Getting ready

We will use only Python code. Make sure you have Python version 3.7 or above. You can use the following command on your CLI to check your version:

```
$ python3 --version
Python 3.8.10
```

The following code execution can be done on a Python shell or a Jupyter notebook.

How to do it...

Let's use the same example we had in the previous, *Inserting logs* recipe, and make some enhancements:

1. Let's start by importing the libraries and defining `basicConfig`. This time, we will set the log level to DEBUG:

    ```
    import csv
    import logging

    logging.basicConfig(filename='our_application.log',
    level=logging.DEBUG)
    ```

2. Then, before declaring the function, we will insert a DEBUG log informing that we are about to test this script:

```
logging.debug(f"Start testing function")
```

3. Next, as we saw in the *Inserting logs* recipe, we will build a function that reads a CSV file and returns the first line but with slight changes. Let's insert a DEBUG message after the first line of the CSV is executed successfully, and a CRITICAL message if we enter the exception:

```
def gets_csv_first_line (csv_file):
    logging.info(f"Starting function to read first line")
    try:
        with open(csv_file, 'r') as file:
            logging.info(f"Reading file")
            reader = csv.reader(file)
            first_row = next(reader)
            logging.debug(f"Finished without problems")
        return first_row
    except Exception as e:
        logging.debug(f"Entered into a exception")
        logging.error(f"Error when reading the CSV file: {e}")
        logging.critical(f"This is a critical error, and the
application needs to stop!")
        raise
```

4. Finally, before we make the call to the function, let's insert a warning message informing that we are about to start it:

```
logging.warning(f"Starting the function to get the first line of
a CSV")
gets_csv_first_line("listings.csv")
```

5. After calling the function, you should see the following output in the our_application.log file:

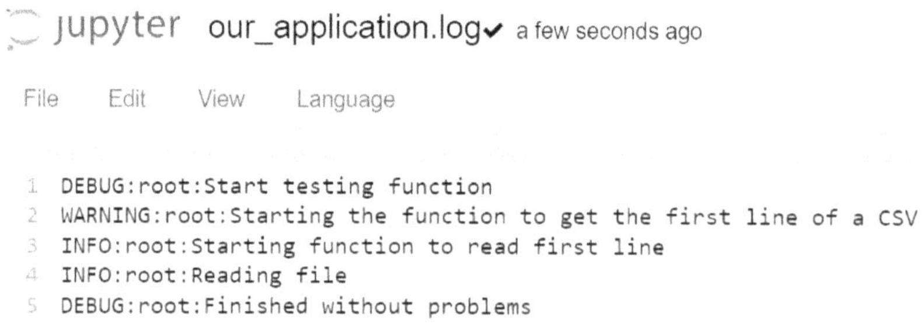

Figure 8.3 – our_application.log updated with different log levels

It informs us the function was correctly executed and no error occurred.

6.   Let's now simulate an error. You should now see the following message inside the `our_ application.log` file:

```
 7   WARNING:root:Starting the function to get the first line of a CSV
 8   INFO:root:Starting function to read first line
 9   INFO:root:Reading file
10   DEBUG:root:Entered into a exception
11   ERROR:root:Error when reading the CSV file: name 'readers' is not defined
12   CRITICAL:root:This is a critical error, and the application needs to stop!
13
```

Figure 8.4 – our_application.log showing an ERROR message

As you can see, we entered the exception, and we can see the `ERROR` and `CRITICAL` messages.

How it works...

Although it may seem irrelevant, we have made beneficial improvements to our function. Each log level corresponds to a different degree of criticality relating to what is happening. Let's take a look at the following figure , which shows the weight of each level:

Numeric values for log levels

Figure 8.5 – Diagram of log level weight according to criticality

Depending on where the log level is inserted, it can prevent the script from continuing and creating a chain of errors since we can add different error handlers according to the level.

By default, the Python logging library is configured to show messages only from the **WARNING** level upward. With this, **DEBUG** and **INFO** messages will not be displayed or saved. That's why, at the beginning of our script, we redefined this initial level to DEBUG, as you can see here:

```
logging.basicConfig(filename='our_application.log', level=logging.
DEBUG)
```

The purpose of showing only **WARNING** messages and above is to avoid spamming the console output or a log file with unnecessary system information. In the following figure, you can see how Python internally organizes its log levels:

Level	When it's used
DEBUG	Detailed information, typically of interest only when diagnosing problems.
INFO	Confirmation that things are working as expected.
WARNING	An indication that something unexpected happened, or indicative of some problem in the near future (e.g. 'disk space low'). The software is still working as expected.
ERROR	Due to a more serious problem, the software has not been able to perform some function.
CRITICAL	A serious error, indicating that the program itself may be unable to continue running.

Figure 8.6 – Log level detailed description and when it's best to use them (source:
`https://docs.python.org/3/howto/logging.html`)

You can use this table as a reference when setting your log messages inside your code. It can also be found in the official Python documentation at `https://docs.python.org/3/howto/logging.html#when-to-use-logging`.

In this recipe, we tried to cover all the log severity levels to demonstrate the recommended places to insert them. Even though it may seem like simple stuff, knowing when each level should be used makes all the difference and brings maturity to your application.

There's more...

Usually, each language has its structured form of logging levels. However, there is an *agreement* in the software engineering world about how the levels should be used. The following figure shows a fantastic decision diagram created by *Taco Jan Osinga* about the behavior of logging levels at the **Operating System (OS)** level.

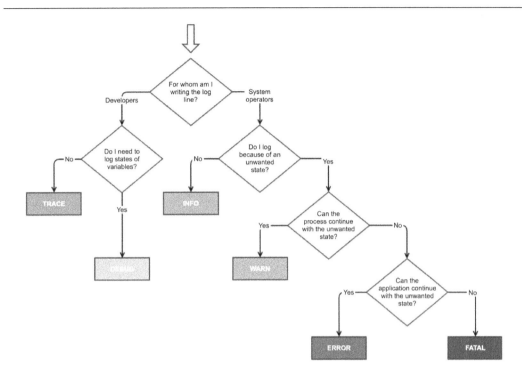

Figure 8.7 – Decision diagram of log levels by Taco Jan Osinga (source: https://
stackoverflow.com/users/3476764/taco-jan-osinga?tab=profile)

See also

For more detailed information about the foundations of Python logs, refer to the official documentation:
`https://docs.python.org/3/howto/logging.html`

Creating standardized logs

Now that we know the best practices for inserting logs and using log levels, we can add more relevant information to our logs to help us monitor our code. Information such as date and time or the module or function executed helps us determine where an issue occurred or where improvements are required.

Creating standardized formatting for application logs or (in our case) data pipeline logs makes the debugging process more manageable, and there are a variety of ways to do this. One way of doing it is to create `.ini` or `.conf` files that hold the configuration on how the logs will be formatted and applied to our wider Python code, for instance.

In this recipe, we will learn how to create a configuration file that will dictate how the logs will be formatted across the code and shown in the execution output.

Getting ready

Let's use the same code as the previous *Using log-level types* recipe, but with more improvements!

You can use the following code to follow the steps of this recipe in a new file or notebook, or reuse the function from the *Using log-level types* recipe. I prefer to make a copy so the first piece of code is left intact:

```
Def gets_csv_first_line(csv_file):
    logger.debug(f"Start testing function")
    logger.info(f"Starting function to read first line")
    try:
        with open(csv_file, 'r') as file:
            logger.info(f"Reading file")
            reader = csv.reader(file)
            first_row = next(reader)
            logger.debug(f"Finished without problems")
        return first_row
    except Exception as e:
        logger.debug(f"Entered into a exception")
        logger.error(f"Error when reading the CSV file: {e}")
        logger.critical(f"This is a critical error, and the
application needs to stop!")
        raise
```

How to do it...

Here are the steps to perform this recipe:

1. To start our exercise, let's create a file called `logging.conf`. My recommendation is to store it in the same location as your Python scripts. However, feel free to keep it somewhere else, but do remember we will need the file's path later.

2. Next, paste the following code inside the `logging.conf` file and save it:

    ```
    [loggers]
    keys=root,data_ingest

    [handlers]
    keys=fileHandler, consoleHandler

    [formatters]
    keys=logFormatter

    [logger_root]
    level=DEBUG
    ```

```
handlers=fileHandler

[logger_data_ingest]
level=DEBUG
handlers=fileHandler, consoleHandler
qualname=data_ingest
propagate=0

[handler_consoleHandler]
class=StreamHandler
level=DEBUG
formatter=logFormatter
args=(sys.stdout,)

[handler_fileHandler]
class=FileHandler
level=DEBUG
formatter=logFormatter
args=('data_ingest.log', 'a')

[formatter_logFormatter]
format=%(asctime)s - %(name)s - %(levelname)s - %(message)s
```

3. Then, insert the following `import` statements, the `config.fileConfig()` method, and the `logger` variable before the `gets_csv_first_line()` function, as you can see here:

```
import csv
import logging
from logging import config

# Loading configuration file
config.fileConfig("logging.conf")

# Creates a log configuration
logger = logging.getLogger("data_ingest")

def gets_csv_first_line(csv_file):
    ...
```

Observe that we are passing `logging.conf` as a parameter for the `config.fileConfig()` method. Pass the whole path if you stored it in a different directory level of your Python script.

4. Now, let's call our function by passing a CSV file. As usual, I will use the `listings.csv` file for this exercise:

```
gets_csv_first_line("listings."sv")
```

You should see the following output in your notebook cell or Python shell:

```
2023-03-19 19:56:23,433 - data_ingest - DEBUG - Start testing function
2023-03-19 19:56:23,434 - data_ingest - INFO - Starting function to read first line
2023-03-19 19:56:23,435 - data_ingest - INFO - Reading file
2023-03-19 19:56:23,436 - data_ingest - DEBUG - Finished without problems
```

Figure 8.8 – Console output with formatted logs

5. Then, check your directory. You should see a file named `data_ingest.log`. Open it, and you should see something like the following screenshot:

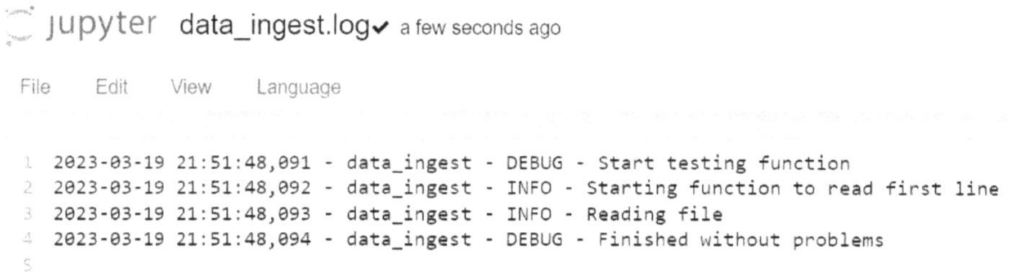

Figure 8.9 – The data_ingest.log file with formatted logs

As you can observe, we created a standardized log format for both console and file output. Let's now understand how we did it in the next section.

How it works...

Before jumping into the code, let's first understand what a configuration file is. **Configuration files**, commonly with a `.conf` or `.ini` file extension, offer a useful way to create customized applications to interact with other applications. You can find some of them inside your OS in the `/etc` or `/var` directories.

Our case is no different. At the beginning of our recipe, we created a configuration file called `logging.conf` that holds the pattern for the Python logs we will apply across our application.

Now, let's take a look inside the `logging.conf` file. Looking closely, it is possible to see some values inside square brackets. Let's start with the first three, as you can see here:

```
[loggers]
[handlers]
[formatters]
```

These parameters are modular components of the Python logging library, allowing easy customization due to their detachment from each other. In short, they represent the following:

- Loggers are used by the code and expose the interface for itself. By default, there is a `root` logger used by Python. For new loggers, we use the `key` argument.

- Handlers send the logs to the configured destination. In our case, we created two: `fileHandler` and `consoleHandler`.

- Formatters create a layout for the log records.

After declaring the basic parameters, we inserted two customized loggers and handlers, as you can observe in the following piece of code:

```
[logger_root]
[logger_data_ingest]
[handler_consoleHandler]
[handler_fileHandler]
```

Creating a customization for the `root Logger` is not mandatory, but here we wanted to change the default log level to `DEBUG` and always send it to `fileHandler`. For `logger_data_ingest`, we also passed `consoleHandler`.

Speaking of handlers, they have a fundamental role here. Although they share the same log level and `Formatter`, they inherit different classes. The `StreamHandler` class catches the log records, and with `args=(sys.stdout,)` it gets all the system outputs for display in the console output. `FileHandler` works similarly, saving all the results at the `DEBUG` level and above.

Finally, `Formatter` dictates how the log will be displayed. There are many ways to set the format, even passing the line of the code where the log was executed. You can see all the possible attributes at `https://docs.python.org/3/library/logging.html#logrecord-attributes`.

The official Python documentation has an excellent diagram, shown in the following figure, that outlines the relationship between these modifiers and another one we didn't cover here, called `Filter`.

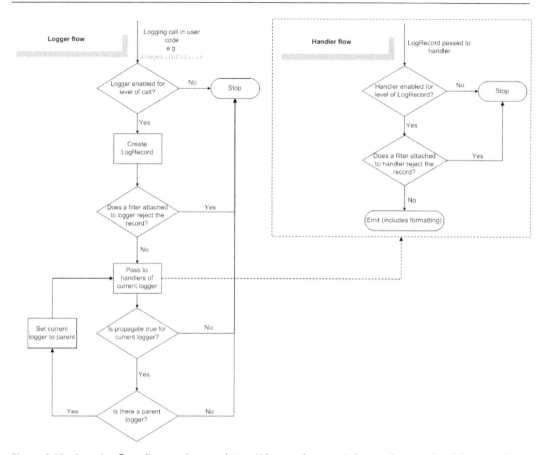

Figure 8.10 – Logging flow diagram (source: https://docs.python.org/3/howto/logging.html#logging-flow)

In our exercise, we created a simple logger handler called `data_ingest` along with the `gets_csv_first_line()` function. Now, imagine how it could be expanded through a whole application or system. Using a single configuration file, we can create several patterns with different applications for different scripts or ETL phases. Let's take a look at the first lines of our code:

```
(...)
config.fileConfig("logging.conf")
logger = logging.getLogger("data_ingest")
(...)
```

`config.fileConfig()` loads the configuration file and `logging.getLogger()` loads the `Logger` instance to use. It will use the `root` as default if it doesn't find the proper `Logger`.

Software engineers commonly use this best practice in a real-world application to avoid code redundancy and create a centralized solution.

There's more...

There are some other acceptable file formats with which to create log configurations. For example, we can use a **YAML Ain't Markup Language** (**YAML**) file or a Python dictionary.

```
version: 1
formatters:
  simple:
    format: '%(asctime)s - %(name)s - %(levelname)s - %(message)s'
handlers:
  console:
    class: logging.StreamHandler
    level: DEBUG
    formatter: simple
    stream: ext://sys.stdout
loggers:
  simpleExample:
    level: DEBUG
    handlers: [console]
    propagate: no
root:
  level: DEBUG
  handlers: [console]
```

Figure 8.11 – Configuration file formatting with YAML (source: https://docs.
python.org/3/howto/logging.html#configuring-logging)

If you want to know more about using the `logging.config` package to create improved YAML or dictionary configurations, refer to the official documentation here: `https://docs.python.org/3/library/logging.config.html#logging-config-api`

See also

To read and understand more about how handlers work, refer to the official documentation here: `https://docs.python.org/3/library/logging.handlers.html`

Monitoring our data ingest file size

When ingesting data, we can track a few items to ensure the incoming information is what we expect. One of the most important of these items is the data size we are ingesting, which can mean file size or the size of chunks of streaming data.

Logging the size of incoming data allows the creation of intelligent and efficient monitoring. If at some point the size of incoming data diverges from what is expected, we can take action to investigate and resolve the issue.

In this recipe, we will create simple Python code that logs the size of ingested files, which is very valuable in data monitoring.

Getting ready

We will use only Python code. Make sure you have Python version 3.7 or above. You can use the following command on your CLI to check your version:

```
$ python3 --version
Python 3.8.10
```

The following code execution can be done using a Python shell or a Jupyter notebook.

How to do it...

This exercise will create a simple Python function to read a file path and return its size in bytes by default. If we want to return the value in megabytes, we only need to pass the input parameter as `True`:

1. Let's start by importing the `os` library:

    ```
    import os
    ```

2. Then, we declare our function that requires a file path, along with an optional parameter to convert the size to megabytes:

    ```
    def get_file_size(file_name, s_megabytes=False):
    ```

3. Let's use `os.stat()` to retrieve information from the file:

    ```
    file_stats = os.stat(file_name)
    ```

4. Since it is optional, we can create an `if` condition to convert the `bytes` value to `megabytes`. If not flagged as `True`, we return the value in `bytes`, as you can see in the following code:

    ```
    if s_megabytes:
        return f"The file size in megabytes is: {file_stats.
    st_size / (1024 * 1024)}"
    return f"The file size in bytes is: {file_stats.st_size}"
    ```

5. Finally, let's call our function, passing a dataset we already used:

    ```
    file_name = "listings.csv"
    get_file_size(file_name)
    ```

    For the `listings.csv` file, you should see the following output:

    ```
    'The file size in bytes is: 965593'
    ```

    Figure 8.12 – File size in bytes

If we execute it by passing s_megabytes as True, we will see the following output:

```
'The file size in megabytes is: 0.9208612442016602'
```

Figure 8.13 – File size in megabytes

Feel free to test it using any file path on your machine and check whether the size is the same as that indicated in the console output.

How it works...

A file's size estimation is convenient when working with data. Let's understand the pieces of code we used to achieve this estimation.

The first operation we used was the os.stat() method to retrieve information about the file, as you can see here:

```
file_stats = os.stat(file_name)
```

This method interacts directly with your OS. If we execute it in isolation, we will have the following output for the listings.csv file:

```
os.stat_result(st_mode=33188, st_ino=302194, st_dev=2064, st_nlink=1,
st_uid=1000, st_gid=1000, st_size=965593, st_atime=1679223955, st_mti
me=1668553769, st_ctime=1676313639)
```

Figure 8.14 – Attributes of the listings.csv file when using os.stat_result

In our case, we only need st_size to bring the bytes estimation, so we called it later in the return clause as follows:

```
file_stats.st_size
```

If you want to know more about the other results shown, you can refer to the official documentation page here: https://docs.python.org/3/library/stat.html#stat.filemode

Lastly, to provide the result in megabytes, we only need to do a simple conversion using the st_size value, where 1 KB is 1,024 bytes, and 1 MB is equal to 1,024 KB. You can see the conversion formula here:

```
file_stats.st_size / (1024 * 1024)
```

There's more...

This recipe showed how easy it is to create a Python function that retrieves the file size. Unfortunately, at the time of writing, there was no straightforward solution to perform the same thing using PySpark.

Glenn Franxman, a software engineer, proposed on his GitHub a workaround solution using Spark internals to estimate the size of a DataFrame. You can see his code on his GitHub at the following link – make sure to give the proper credits if you do use it: `https://gist.github.com/ gfranxman/4fd0719ff2618039182dd7ea1a702f8e`

Let's use Glenn's code in an example to estimate the DataFrame size and see how it works:

```
from pyspark.serializers import PickleSerializer,
AutoBatchedSerializer
def _to_java_object_rdd(rdd):
    """ Return a JavaRDD of Object by unpickling
    It will convert each Python object into Java object by Pyrolite,
whenever the
    RDD is serialized in batch or not.
    """
    rdd = rdd._reserialize(AutoBatchedSerializer(PickleSerializer()))
    return rdd.ctx._jvm.org.apache.spark.mllib.api.python.SerDe.
pythonToJava(rdd._jrdd, True)

def estimate_df_size(df):
    JavaObj = _to_java_object_rdd(df.rdd)
    nbytes = spark._jvm.org.apache.spark.util.SizeEstimator.
estimate(JavaObj)
    return nbytes
```

To execute the preceding code, you must have a SparkSession initiated. Once you have this and a DataFrame, execute the code and call the `estimate_df_size()` function as follows:

```
estimate_df_size(df)
```

You should see the following output in bytes, depending on which DataFrame you are using:

13029456

Figure 8.15 – DataFrame size in bytes

Remember that this solution will only work if you pass a DataFrame as a parameter. Our Python code works well for other file estimations and doesn't have performance issues when estimating big files.

See also

Unlike PySpark, Scala has a `SizeEstimator` function to return the size of a DataFrame. You can find more here: `https://spark.apache.org/docs/latest/api/java/index. html?org/apache/spark/util/SizeEstimator.html`

Logging based on data

As mentioned in the *Monitoring our data ingest file size* recipe, logging our ingest is a good practice in the data field. There are several ways to explore our ingestion logs to increase the process's reliability and our confidence in it. In this recipe, we will start to get into the data operations field (or **DataOps**), where the goal is to track the behavior of data from the source until it reaches its final destination.

This recipe will explore other metrics we can track to create a reliable data pipeline.

Getting ready

For this exercise, let's imagine we have two simple data ingests, one from a database and another from an API. Since this is a straightforward pipeline, let's visualize it with the following diagram:

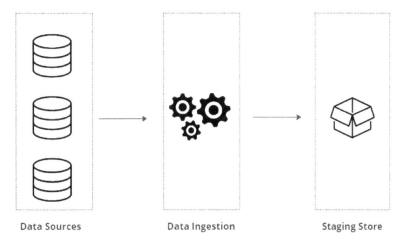

Figure 8.16 – Data ingestion phases

With this in mind, let's explore the instances we can log to make monitoring efficient.

How to do it...

Let's define the essential metrics based on each layer (or step) we saw in the preceding diagram:

1. **Data sources**: Let's start with the first layer of the ingestion—the sources. Knowing we are handling two different data sources, we must create additional metrics for them. See the following figure for reference:

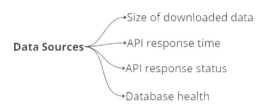

Figure 8.17 – Database metrics to monitor

2. **Ingestion**: Now that the source is logged and monitored, let's move on to the ingestion layer. As we saw previously in this chapter, we can log information such as errors, informative parts of the code execution, file size, and so on. Let's insert more content here, such as the schema and the time taken to retrieve or process data. We will end up with a diagram similar to the following:

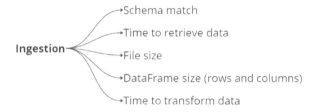

Figure 8.18 – Data ingestion metrics to monitor

3. **Staging layer**: Lastly, let's cover the final layer after ingestion. The goal here is to ensure we maintain the integrity of the data, so verifying whether the schema still matches the data is crucial. We can also add logs to monitor the number of Parquet files and their sizes. See the following figure for reference:

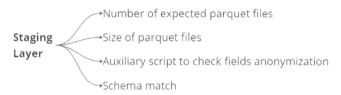

Figure 8.19 – Staging layer metrics to monitor

Now that we have covered the essential topics to be monitored, let's understand why they were chosen.

How it works...

Since the first recipe of this chapter, we have perpetually reinforced how logs are relevant to getting your system to work correctly. Here, we put it all together to see, albeit from a high level, how storing specific information can help us with monitoring.

Starting with the data source layer, the metrics chosen were based on the response of the data and the availability to retrieve data. Understanding whether we can begin the ingestion process is fundamental, and even more important is knowing whether the data size is what we expect.

Imagine the following scenario: every day, we ingest 50 MB of data from an API. One day, however, we received 10 KB. With proper logging and monitoring functionalities, we can quickly review the issue in terms of historic events. We can expand the data size check to the subsequent layers we covered in the recipe.

> **Note**
>
> We purposely intercalate the words "step" and "layer" when referring to the phases of the ingestion process since it can vary in different works of literature and in different companies' internal processes.

Another way to log and monitor our data is by using schema validation. **Schema validation** (when applicable) guarantees that nothing has changed at the source. Therefore, the results for transformation or aggregation tend to be linear. We can also implement an auxiliary function or job to check that fields containing **sensitive** or **Personally Identifiable Information** (**PII**) are adequately anonymized.

Monitoring the *parquet file size or count* is crucial to verify that quality is being maintained. As seen in *Chapter 7*, the number of parquet files can interfere with other applications' reading quality or even the ETL's subsequent phases.

Finally, it is essential to point out that we covered logs here to ensure the quality and reliability of our data ingestion. Remember that the best practice is to align the records we got from the code with the examples we saw here.

There's more...

The content of this recipe is part of a more extensive subject called **data observability**. Data observability is a union of data operations, quality, and governance. The objective is to centralize everything to make the management and monitoring of data processes efficient and reliable, bringing health to data.

We will discuss this further in *Chapter 12*. However, if you are curious about the topic, Databand (an IBM company) has a good introduction at `https://databand.ai/data-observability/`.

See also

Find out more about monitoring ETL pipelines at the DataGaps blog page, here: `https://www.datagaps.com/blog/monitoring-your-etl-test-data-pipelines-in-production-dataops-suite/`

Retrieving SparkSession metrics

Until now, we created our logs to provide more information and be more useful for monitoring. Logging allows us to build customized metrics based on the necessity of our pipeline and code. However, we can also take advantage of built-in metrics from frameworks and programming languages.

When we create a `SparkSession`, it provides a web UI with useful metrics that can be used to monitor our pipelines. Using this, the following recipe shows you how to access and retrieve metric information from SparkSession, and use it as a tool when ingesting or processing a DataFrame.

Getting ready

You can execute this recipe using the PySpark command line or the Jupyter Notebook.

Before exploring the Spark UI metrics, let's create a simple `SparkSession` using the following code:

```
from pyspark.sql import SparkSession
spark = SparkSession.builder \
        .master("local[1]") \
        .appName("chapter8_monitoring") \
        .config("spark.executor.memory", '3g') \
        .config("spark.executor.cores", '3') \
        .config("spark.cores.max", '3') \
        .getOrCreate()
```

Then, let's read a JSON file and call the `.show()` method as follows:

```
df_json = spark.read.option("multiline","true") \
                    .json('github_events.json')
df_json.show()
```

I am using a dataset called `github_events.json`, which we worked with previously in *Chapter 4*. However, feel free to use whatever you prefer, since the objective here is not to observe the schema of the dataset, but to see what we can find out from the Spark UI.

How to do it...

1. With the `SparkSession` initiated as outlined in the *Getting ready* section, we can use the `spark` command to retrieve a link to the Spark UI, as follows:

    ```
    spark
    ```

You should see the following output:

SparkSession - in-memory
SparkContext

Spark UI
Version
v3.1.2
Master
local[1]
AppName
chapter8_monitoring

Figure 8.20 – spark command output

2. Click on **Spark UI**, and your browser will open a new tab. You should see a page like this:

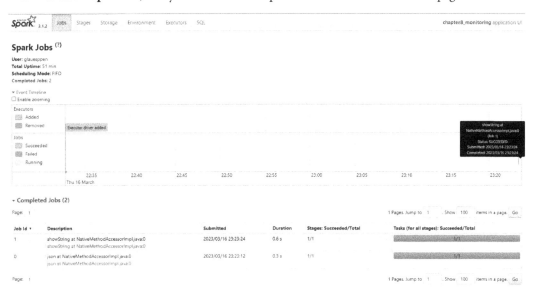

Figure 8.21 – Spark UI: Jobs page view

If you are not using the Jupyter Notebook, you can access this interface by pointing your browser to http://localhost:4040/.

My page looks more crowded because I expanded **Event Timeline** and **Completed Jobs** – you can do the same by clicking on them.

3.   Next, let's explore the first completed job further. Click on **showString at NativeMethodAccessorImpl.java:0** and you should see the following page:

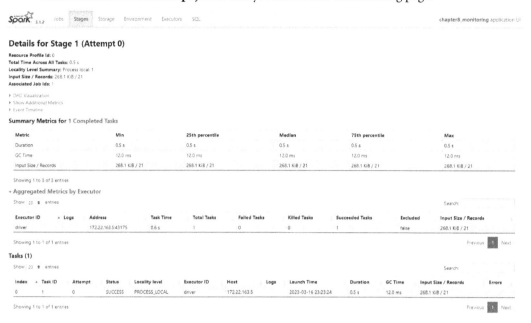

Figure 8.22 – Spark UI: Stage page view for a specific job

Here, we can see the task status of this job in more detail, covering things such as how much memory it used, the time taken to execute it, and so on.

Note also that it switched to the **Stages** tab at the top menu.

4.   Now, click on the **Executors** button at the top of the page. You should see a page similar to this:

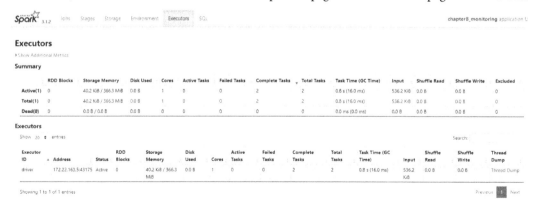

Figure 8.23 – Spark UI: Executors page view

All the metrics here are related to the Spark drivers and nodes.

5.   Then, click on the **SQL** button in the top menu. You should see the following page:

Figure 8.24 – Spark UI: SQL page view

On this page, it is possible to see the queries executed by Spark internally. If we used an explicit query in our code, we would see here how it was performed internally.

You don't need to worry about the **Details** menu being expanded. The objective is to show that it is possible to track Spark's actions when executing the `.show()` method.

How it works...

Now that we have explored Spark UI, let's understand how each tab is organized and some of the steps we did with them.

In *step 2*, we had a first glance at the interface. This interface makes it possible to see an event timeline with information about when the driver was created and executed. Also, we can observe the jobs marked on the timeline, as follows:

Figure 8.25 – Detailed view of the Event Timeline expanded menu

We can observe how they interact when working with bigger jobs and more complex parallel tasks. Unfortunately, we would need a dedicated project and several datasets to simulate this, but you now know where to look for future reference.

Then, we selected **showString at NativeMethodAccessorImpl.java:0**, which redirected us to the **Stages** page. This page offers more detailed information about Spark's tasks, whether the task was successful or not.

An excellent metric and visualization tool is **DAG Visualization** (referring to directed acyclic graphs), which can be expanded and will show something like this:

Details for Stage 1 (Attempt 0)

Resource Profile Id: 0
Total Time Across All Tasks: 0.5 s
Locality Level Summary: Process local: 1
Input Size / Records: 268.1 KiB / 21
Associated Job Ids: 1

▼ DAG Visualization

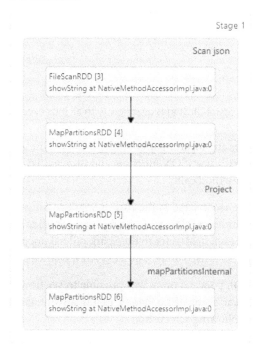

Figure 8.26 – DAG Visualization of a job

This offers an excellent overview of each step performed at each stage. We can also consult this to understand which part was problematic in the event of an error based on the traceback message.

Since we selected a specific task (or job), it showed its stages and details. However, we can display all the steps executed if we go directly to **Stages**. Doing so, you should see something like this:

Figure 8.27 – Spark UI: Stages overview with all jobs executed

Although the description messages are not straightforward, we can get the gist of what each of them is doing. `Stage id 0` refers to the reading JSON function, and `Stage id 1` with the `showString` message refers to the `.show()` method call.

The **Executors** page shows the metrics related to the core of Spark and how it is performing. You can use this information to understand your cluster's behavior and whether any tuning is needed. For more detailed information about each field, refer to the Spark official documentation at `https://spark.apache.org/docs/latest/monitoring.html#executor-metrics`.

Last but not least, we saw the **SQL** page, where it was possible to see how Spark internally shuffles and aggregates the data behind the scenes, like **Stages**, taking advantage of a more visual form of execution, as you can see in the following screenshot:

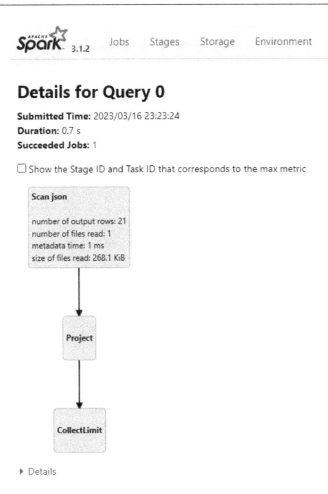

Figure 8.28 – Flow diagram of the SQL query internally executed

Here, we can see that the query is related to the `.show()` method. There is helpful information inside it, including the number of output rows, the files read, and their sizes.

There's more...

Even though Spark metrics are handy, you might wonder how to use them when hosting your PySpark jobs on cloud providers such as AWS or Google Cloud.

AWS provides a simple solution to enable Spark UI when using **AWS Glue**. You can find out more about it at `https://docs.aws.amazon.com/glue/latest/dg/monitor-spark-ui-jobs.html`.

Google Data Proc provides a web interface for its cluster, where you can also see metrics for **Hadoop** and **YARN**. Since Spark runs on top of YARN, you won't find a link directly for Spark UI, but you can use the YARN interface to access it. You can find out more here: `https://cloud.google.com/dataproc/docs/concepts/accessing/cluster-web-interfaces`.

See also

Towards Data Science has a fantastic article about Spark metrics: `https://towardsdatascience.com/monitoring-of-spark-applications-3ca0c271c4e0`

Further reading

- `https://spark.apache.org/docs/latest/monitoring.html#executor-task-metrics`

- `https://developer.here.com/documentation/metrics-and-logs/user_guide/topics/spark-ui.html`

- `https://github.com/LucaCanali/Miscellaneous/blob/master/Spark_Notes/Spark_TaskMetrics.md`

- `https://docs.python.org/3/howto/logging.html`

- `datadoghq.com/blog/python-logging-best-practices/`

- `https://coralogix.com/blog/python-logging-best-practices-tips/`

9

Putting Everything Together with Airflow

So far, we have covered the different aspects and steps of data ingestion. We have seen how to configure and ingest structured and unstructured data, what analytical data is, and how to improve logs for more insightful monitoring and error handling. Now is the time to group all this information to create something similar to a real-world project.

From now on, in the following chapters, we will use Apache Airflow, an open source platform that allows us to create, schedule, and monitor workflows. Let's start our journey by configuring and understanding the fundamental concepts of Apache Airflow and how powerful this tool is.

In this chapter, you will learn about the following topics:

- Configuring Airflow
- Creating DAGs
- Creating custom operators
- Configuring sensors
- Creating connectors in Airflow
- Creating parallel ingest tasks
- Defining ingest-dependent DAGs

By the end of this chapter, you will have learned about the most important components of Airflow and how to configure them, including how to solve related issues in this process.

Technical requirements

You can find the code for this chapter in the GitHub repository here: `https://github.com/PacktPublishing/Data-Ingestion-with-Python-Cookbook`

Installing Airflow

This chapter requires that Airflow is installed on your local machine. You can install it directly on your **Operating System** (**OS**) or using a Docker image. For more information regarding this, refer to the *Configuring Docker for Airflow* recipe in *Chapter 1*.

Configuring Airflow

Apache Airflow has many capabilities and a quick setup, which helps us start designing our workflows as code. Some additional configurations might be required as we progress with the workflows and into data processing. Gladly, Airflow has a dedicated file for inserting other arrangements without changing anything within its core.

In this recipe, we will learn more about the `airflow.conf` file, how to use it, and other valuable configurations required to execute the other recipes in this chapter. We will also cover where to find this file and how other environment variables work with this tool. Understanding these concepts in practice helps us to identify potential improvements or solve problems.

Getting ready

Before moving on to the code, ensure your Airflow runs correctly. You can do that by checking the Airflow UI at this link: `http://localhost:8080`.

If you are using a Docker container (as I am) to host your Airflow application, you can check its status on the terminal with the following command:

```
$ docker ps
```

This is the output:

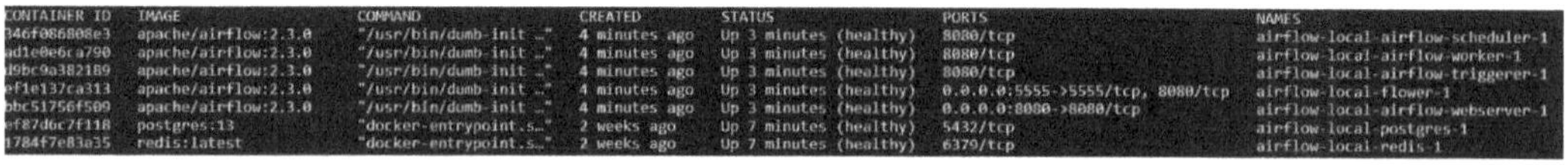

CONTAINER ID	IMAGE	COMMAND	CREATED	STATUS	PORTS	NAMES
346f086808e3	apache/airflow:2.3.0	"/usr/bin/dumb-init …"	4 minutes ago	Up 3 minutes (healthy)	8080/tcp	airflow-local-airflow-scheduler-1
ad1e0e6ca790	apache/airflow:2.3.0	"/usr/bin/dumb-init …"	4 minutes ago	Up 3 minutes (healthy)	8080/tcp	airflow-local-airflow-worker-1
d9bc9a382189	apache/airflow:2.3.0	"/usr/bin/dumb-init …"	4 minutes ago	Up 3 minutes (healthy)	8080/tcp	airflow-local-airflow-triggerer-1
ef1e137ca313	apache/airflow:2.3.0	"/usr/bin/dumb-init …"	4 minutes ago	Up 3 minutes (healthy)	0.0.0.0:5555->5555/tcp, 8080/tcp	airflow-local-flower-1
bbc51756f509	apache/airflow:2.3.0	"/usr/bin/dumb-init …"	4 minutes ago	Up 3 minutes (healthy)	0.0.0.0:8080->8080/tcp	airflow-local-airflow-webserver-1
ef87d6c7f118	postgres:13	"docker-entrypoint.s…"	2 weeks ago	Up 7 minutes (healthy)	5432/tcp	airflow-local-postgres-1
1784f7e83a35	redis:latest	"docker-entrypoint.s…"	2 weeks ago	Up 7 minutes (healthy)	6379/tcp	airflow-local-redis-1

Figure 9.1 – Airflow Docker containers running

Or, you can check the container status on Docker Desktop, as in the following screenshot:

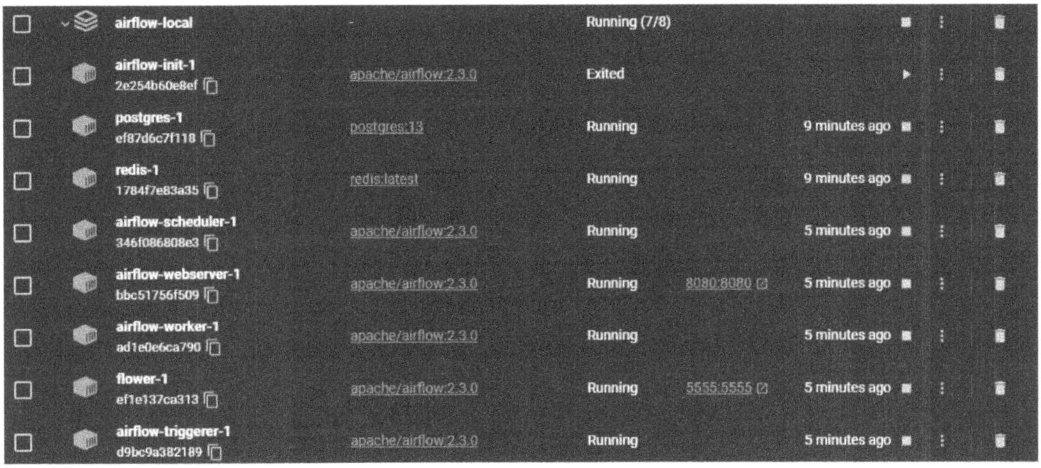

Figure 9.2 – Docker Desktop view of Airflow containers running

How to do it...

Here are the steps to perform this recipe:

1. Let's start by installing the MongoDB additional provider for Airflow. If you are using the `docker-compose.yaml` file, open it and add `apache-airflow-providers-mongo` inside `_PIP_ADDITIONAL_REQUIREMENTS`. Your code will look like this:

```
version: '3'
x-airflow-common:
  &airflow-common
  # In order to add custom dependencies or upgrade provider packages you can use your extended image.
  # Comment the image line, place your Dockerfile in the directory where you placed the docker-compose
  # and uncomment the "build" line below, Then run `docker-compose build` to build the images.
  image: ${AIRFLOW_IMAGE_NAME:-apache/airflow:2.3.0}
  # build: .
  environment:
    &airflow-common-env
    AIRFLOW__CORE__EXECUTOR: CeleryExecutor
    AIRFLOW__DATABASE__SQL_ALCHEMY_CONN: postgresql+psycopg2://airflow:airflow@postgres/airflow
    # For backward compatibility, with Airflow <2.3
    AIRFLOW__CORE__SQL_ALCHEMY_CONN: postgresql+psycopg2://airflow:airflow@postgres/airflow
    AIRFLOW__CELERY__RESULT_BACKEND: db+postgresql://airflow:airflow@postgres/airflow
    AIRFLOW__CELERY__BROKER_URL: redis://:@redis:6379/0
    AIRFLOW__CORE__FERNET_KEY: ''
    AIRFLOW__CORE__DAGS_ARE_PAUSED_AT_CREATION: 'true'
    AIRFLOW__CORE__LOAD_EXAMPLES: 'true'
    AIRFLOW__API__AUTH_BACKENDS: 'airflow.api.auth.backend.basic_auth'
    _PIP_ADDITIONAL_REQUIREMENTS: ${_PIP_ADDITIONAL_REQUIREMENTS:-apache-airflow-providers-mongo}
```

Figure 9.3 – The docker-compose.yaml file in the environment variables section

If you are hosting Airflow directly on your machine, you can do the same installation using **PyPi**: `https://pypi.org/project/apache-airflow-providers-mongo/`.

2. Next, we will create a folder called `files_to_test`, and inside it, create two more folders: `output_files` and `sensors_files`. You don't need to worry about its usage yet since it will be used later in this chapter. Your Airflow folder structure should look like this:

```
.
└── your_airflow_folder/
    ├── dags
    ├── plugins
    ├── logs
    ├── files_to_test
    ├── .env
```

Figure 9.4 – Airflow local directory folder structure

3. Now, let's mount the volumes of our Docker image. You can skip this part if you are not using Docker to host Airflow.

In your `docker-compose.yaml` file, under the `volume` parameter, add the following:

```
- ./config/airflow.cfg:/usr/local/airflow/airflow.cfg
- ./files_to_test:/opt/airflow/files_to_test
```

Your final `volumes` section will look like this:

```
volumes:
    - ./dags:/opt/airflow/dags
    - ./logs:/opt/airflow/logs
    - ./plugins:/opt/airflow/plugins
    - ./files_to_test:/opt/airflow/files_to_test
    - ./config/airflow.cfg:/usr/local/airflow/airflow.cfg
```

Figure 9.5 – docker-compose.yaml volumes

Stop and restart your container so these changes can be propagated.

4. Finally, we will fix a bug in the `docker-compose.yaml` file. This official fix for this bug is within the Airflow official documentation and therefore wasn't included in the Docker image. You can see the complete issue and the solution here: `https://github.com/apache/airflow/discussions/24809`.

To fix the bug, go to the `airflow-init` section of the `docker-compose` file and insert `_PIP_ADDITIONAL_REQUIREMENTS: ''` inside the `environment` parameter. Your code will look like this:

```
# yamllint enable rule:line-length
environment:
  <<: *airflow-common-env
  _AIRFLOW_DB_UPGRADE: 'true'
  _AIRFLOW_WWW_USER_CREATE: 'true'
  _AIRFLOW_WWW_USER_USERNAME: ${_AIRFLOW_WWW_USER_USERNAME:-airflow}
  _AIRFLOW_WWW_USER_PASSWORD: ${_AIRFLOW_WWW_USER_PASSWORD:-airflow}
  _PIP_ADDITIONAL_REQUIREMENTS: ''
```

Figure 9.6 – The docker-compose.yaml environment variables
section with PIP_ADDITIONAL_REQUIREMENTS

This action will fix the following issue registered on GitHub: `https://github.com/apache/airflow/pull/23517`.

How it works...

The configuration presented here is simple. However, it guarantees the application will keep working through the chapter recipes.

Let's start with the package we installed in *step 1*. Like other frameworks or platforms, Airflow has its *batteries* included, which means it already comes with various packages. But, as its popularity started to increase, it started to require other types of connections or operators, which the open source community took care of.

You can find a list of released packages that can be installed on Airflow here: `https://airflow.apache.org/docs/apache-airflow-providers/packages-ref.html`.

Before jumping into other code explanations, let's understand the `volume` section inside the `docker-compose.yaml` file. This configuration allows Airflow to see which folders reflect the same respective ones inside the Docker container without the necessity to upload code using a Docker command every time. In other words, we can synchronously add our **Directed Acyclic Graph** (**DAG**) files and new operators and see some logs, among other things, and this will be reflected inside the container.

Next, we declared the Docker mount volume configurations for two parts: the new folder we created (`files_to_test`) and the `airflow.cfg` file. The first one will allow Airflow to replicate the `files_to_test` local folder inside the container, so we can use it to use files in a more simplified way. Otherwise, if we try to use it without the mounting volume, the following error will appear when trying to retrieve any file:

```
2023-03-24, 19:57:11 UTC] {taskinstance.py:1889} ERROR - Task failed with exception
raceback (most recent call last):
 File "/home/airflow/.local/lib/python3.7/site-packages/airflow/operators/python.py", line 171, in execute
   return_value = self.execute_callable()
 File "/home/airflow/.local/lib/python3.7/site-packages/airflow/operators/python.py", line 189, in execute_callable
   return self.python_callable(*self.op_args, **self.op_kwargs)
 File "/opt/airflow/dags/ids_ingest/ids_ingest_dag.py", line 17, in get_ids_from_json
   with open (filename_json, 'r') as f:
ileNotFoundError: [Errno 2] No such file or directory: '/opt/airflow/files_to_test/github_events.json'
```

Figure 9.7 – Error in Airflow when the folder is not referred to in the container volume

Although we will not use the `airflow.cfg` file, for now, it is a good practice to know how to access this file and what it is used for. This file contains the Airflow configurations and can be edited to include more. Usually, sensitive data is stored inside it to prevent other people from having improper access since, by default, the content of the `airflow.cfg` file cannot be accessed in the UI.

> **Note**
>
> Be very cautious when changing or handling the `airflow.cfg` file. This file contains all the required configurations and other relevant settings to make Airflow work. We will explore more about this in *Chapter 10*.

See also

For more information about the Docker image, see the documentation page here: `https://airflow.apache.org/docs/apache-airflow/stable/howto/docker-compose/index.html`.

Creating DAGs

The core concept of Airflow is based on DAGs, which collect, group, and organize tasks to be executed in a specific order. A DAG is also responsible for managing the dependencies between its tasks. Simply put, it is not concerned about what a task is doing but just *how* to execute it. Typically, a DAG starts at a scheduled time, but we can also define dependencies between other DAGs so that they will start based on their execution statuses.

We will create our first DAG in this recipe and set it to run based on a specific schedule. With this first step, we enter into practically designing our first workflow.

Getting ready

Please refer to the *Getting ready* section in the *Configuring Airflow* recipe for this recipe since we will handle it with the same technology.

Also, let's create a directory called `ids_ingest` inside our `dags` folder. Inside the `ids_ingest` folder, we will create two files: `__init__.py` and `ids_ingest_dag.py`. The final structure will look as follows:

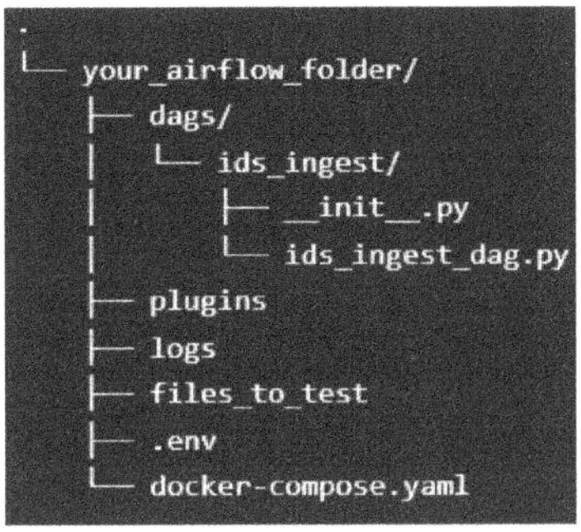

```
.
└── your_airflow_folder/
    ├── dags/
    │   └── ids_ingest/
    │       ├── __init__.py
    │       └── ids_ingest_dag.py
    ├── plugins
    ├── logs
    ├── files_to_test
    ├── .env
    └── docker-compose.yaml
```

Figure 9.8 – Airflow's local directory structure with the ids_ingest DAG

How to do it...

In this exercise, we will write a DAG that retrieves the IDs of the `github_events.json` file. Open `ids_ingest_dag.py`, and let's add the content to write our first DAG:

1. Let's start by importing the libraries we will use in this script. I like to separate the imports from the Airflow library and Python's library as a good practice:

```python
from airflow import DAG
from airflow.settings import AIRFLOW_HOME
from airflow.operators.bash import BashOperator
from airflow.operators.python_operator import PythonOperator

import json
from datetime import datetime, timedelta
```

2. Then, we will define `default_args` for our DAG, as you can see here:

```python
# Define default arguments
default_args = {
    'owner': 'airflow',
    'depends_on_past': False,
    'start_date': datetime(2023, 3, 22),
    'retries': 1,
    'retry_delay': timedelta(minutes=5)
}
```

3. Now, we will create a Python function that receives the JSON file and returns the IDs inside it. Since it is a small function, we can create it inside the DAG's file:

```
def get_ids_from_json(filename_json):
    with open (filename_json, 'r') as f:
        git = json.loads(f.read())
    print([item['id'] for item in git])
```

4. Next, we will instantiate our DAG object, and inside it, we will define two operators: a `BashOperator` instance to show a console message and `PythonOperator` to execute the function we just created, as you can see here:

```
# Instantiate a DAG object
with DAG(
    dag_id='simple_ids_ingest',
    default_args=default_args,
    schedule_interval=timedelta(days=1),
) as dag:

    first_task = BashOperator(
            task_id="first_task",
            bash_command="echo $AIRFLOW_HOME",
        )

    filename_json = f"{AIRFLOW_HOME}/files_to_test/github_
events.json"
    get_id_from_json = PythonOperator(
        task_id="get_id_from_json",
        python_callable=get_ids_from_json,
        op_args=[filename_json]
    )
```

Make sure you save the file before jumping to the next step.

5. Now, head over to the Airflow UI. Although plenty of DAG examples are provided by the Airflow team, you should look for a DAG called `simple_ids_ingest`. You will notice the DAG is not enabled. Click on the toggle button to enable it, and you should have something like the following:

Figure 9.9 – The Airflow DAG enabled on the UI

6. As soon as you enable it, the DAG will start running. Click on the DAG name to be redirected to the DAG's page, as you can see in the following screenshot:

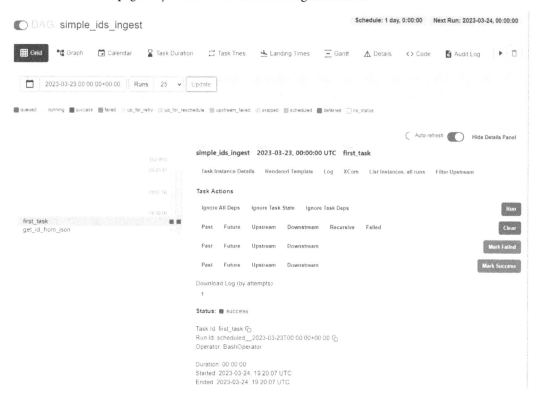

Figure 9.10 – DAG Grid page view

If everything is well configured, your page should look like this:

Figure 9.11 – DAG running successfully in Graph page view

7. Then, click on the **Graph** item in the top menu and click on the get_id_from_json task. A small window will show up as follows:

Figure 9.12 – Task options

8. Then, click the **Log** button. You will be redirected to a new page with the logs for this task, as seen here:

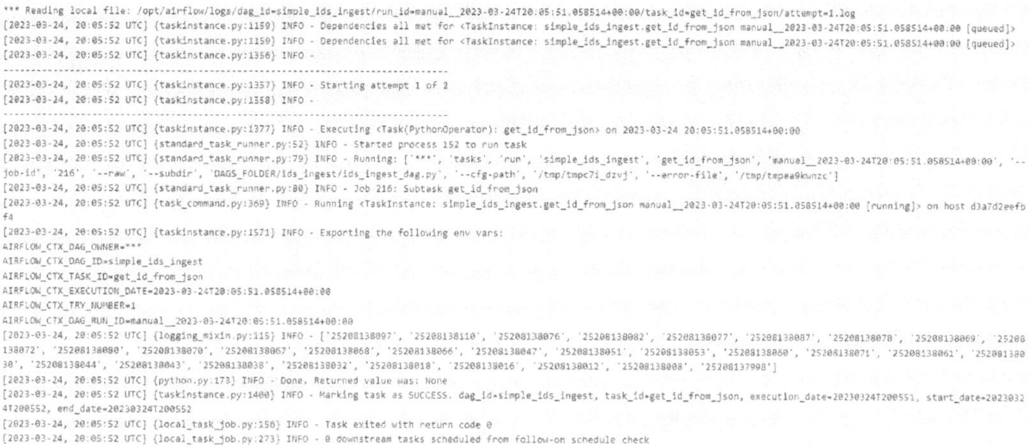

```
*** Reading local file: /opt/airflow/logs/dag_id=simple_ids_ingest/run_id=manual__2023-03-24T20:05:51.058514+00:00/task_id=get_id_from_json/attempt=1.log
[2023-03-24, 20:05:52 UTC] {taskinstance.py:1159} INFO - Dependencies all met for <TaskInstance: simple_ids_ingest.get_id_from_json manual__2023-03-24T20:05:51.058514+00:00 [queued]>
[2023-03-24, 20:05:52 UTC] {taskinstance.py:1159} INFO - Dependencies all met for <TaskInstance: simple_ids_ingest.get_id_from_json manual__2023-03-24T20:05:51.058514+00:00 [queued]>
[2023-03-24, 20:05:52 UTC] {taskinstance.py:1356} INFO -
--------------------------------------------------------------------
[2023-03-24, 20:05:52 UTC] {taskinstance.py:1357} INFO - Starting attempt 1 of 2
[2023-03-24, 20:05:52 UTC] {taskinstance.py:1358} INFO -
--------------------------------------------------------------------
[2023-03-24, 20:05:52 UTC] {taskinstance.py:1377} INFO - Executing <Task(PythonOperator): get_id_from_json> on 2023-03-24 20:05:51.058514+00:00
[2023-03-24, 20:05:52 UTC] {standard_task_runner.py:52} INFO - Started process 152 to run task
[2023-03-24, 20:05:52 UTC] {standard_task_runner.py:79} INFO - Running: ['***', 'tasks', 'run', 'simple_ids_ingest', 'get_id_from_json', 'manual__2023-03-24T20:05:51.058514+00:00', '--
job-id', '216', '--raw', '--subdir', 'DAGS_FOLDER/ids_ingest/ids_ingest_dag.py', '--cfg-path', '/tmp/tmpc7i_dzvj', '--error-file', '/tmp/tmpea9kunzc']
[2023-03-24, 20:05:52 UTC] {standard_task_runner.py:80} INFO - Job 216: Subtask get_id_from_json
[2023-03-24, 20:05:52 UTC] {task_command.py:369} INFO - Running <TaskInstance: simple_ids_ingest.get_id_from_json manual__2023-03-24T20:05:51.058514+00:00 [running]> on host d3a7d2eefb
f4
[2023-03-24, 20:05:52 UTC] {taskinstance.py:1571} INFO - Exporting the following env vars:
AIRFLOW_CTX_DAG_OWNER=***
AIRFLOW_CTX_DAG_ID=simple_ids_ingest
AIRFLOW_CTX_TASK_ID=get_id_from_json
AIRFLOW_CTX_EXECUTION_DATE=2023-03-24T20:05:51.058514+00:00
AIRFLOW_CTX_TRY_NUMBER=1
AIRFLOW_CTX_DAG_RUN_ID=manual__2023-03-24T20:05:51.058514+00:00
[2023-03-24, 20:05:52 UTC] {logging_mixin.py:115} INFO - ['25208138097', '25208138110', '25208138076', '25208138082', '25208138077', '25208138087', '25208138070', '25208138069', '25208
138072', '25208138080', '25208138070', '25208138067', '25208138066', '25208138047', '25208138051', '25208138053', '25208138060', '25208138071', '25208138061', '25208138
30', '25208138044', '25208138043', '25208138038', '25208138032', '25208138010', '25208138016', '25208138012', '25208138008', '25208137998']
[2023-03-24, 20:05:52 UTC] {python.py:173} INFO - Done. Returned value was: None
[2023-03-24, 20:05:52 UTC] {taskinstance.py:1400} INFO - Marking task as SUCCESS. dag_id=simple_ids_ingest, task_id=get_id_from_json, execution_date=20230324T200551, start_date=2023032
4T200552, end_date=20230324T200552
[2023-03-24, 20:05:52 UTC] {local_task_job.py:156} INFO - Task exited with return code 0
[2023-03-24, 20:05:52 UTC] {local_task_job.py:273} INFO - 0 downstream tasks scheduled from follow-on schedule check
```

Figure 9.13 – Task logs in the Airflow UI

As we can see in the preceding screenshot, our task successfully finished and returned the IDs as we expected. You can see the results in the INFO log under the AIRFLOW_CTX_DAG_RUN message.

How it works...

We created our first DAG with a few lines to retrieve and show a list of IDs from a JSON file. Now, let's understand how it works.

To start with, we created our files under the dags directory. It happens because, by default, Airflow will understand everything inside of it as a DAG file. The folder we created inside of it was just for organization purposes, and Airflow will ignore it. Along with the ids_ingest_dag.py file, we also made an __init__.py file. This file internally tells Airflow to look inside this folder. As a result, you will see the following structure:

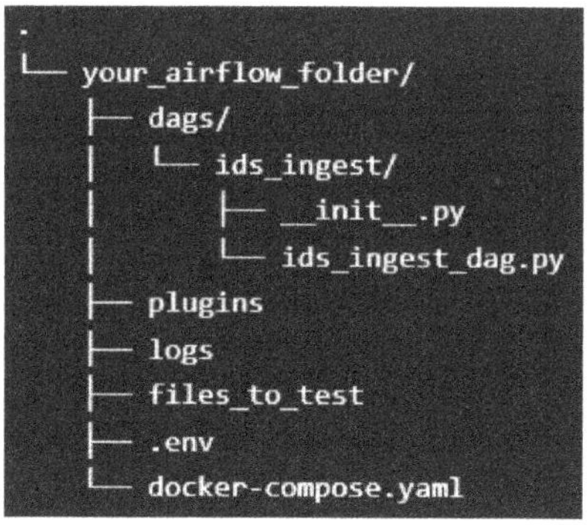

```
.
└── your_airflow_folder/
    ├── dags/
    │   └── ids_ingest/
    │       ├── __init__.py
    │       └── ids_ingest_dag.py
    ├── plugins
    ├── logs
    ├── files_to_test
    ├── .env
    └── docker-compose.yaml
```

Figure 9.14 – Airflow local directory structure with the ids_ingest DAG

> **Note**
>
> As you might be wondering, it is possible to change this configuration, but I don't recommend this at all since other internal packages might depend on it. Do it only in the case of extreme necessity.

Now, let's take a look at our instantiated DAG:

```
with DAG(
    dag_id='simple_ids_ingest',
    default_args=default_args,
    schedule_interval=timedelta(days=1),
) as dag:
```

As you can observe, creating a DAG is simple, and its parameters are spontaneous. `dag_id` is crucial and must be unique; otherwise, it can create confusion and merge with other DAGs. The `default_args` we declared in *step 2* will guide the DAG, telling when it needs to be executed, its user owner, the number of retries in case of a failure, and other valuable parameters. After the `as dag` declaration, we inserted the bash and Python operators, and they must be indented to be understood as the DAG's tasks.

Finally, to set our workflow, we declared the following line:

```
first_task >> get_id_from_json
```

As we might guess, it sets the order of which task should be executed first.

There's more...

We saw how easy it is to create a task to execute a Python function and a bash command. By default, Airflow comes with some handy operators to be used daily within a data ingestion pipeline. For more information, you can refer to the official documentation page here: `https://airflow.apache.org/docs/apache-airflow/stable/_api/airflow/operators/index.html`.

Tasks, operators, XCom, and others

Airflow DAGs are a powerful way to group and execute operations. Besides the task and operators we saw here, DAGs support other types of workloads and communication across other tasks or DAGs. Unfortunately, since that is not the main subject of this book, we will not cover those concepts in detail, but I highly recommend reading the official documentation here: `https://airflow.apache.org/docs/apache-airflow/stable/core-concepts/index.html`.

Error handling

If you encounter any errors while building this DAG, you can use the instructions from *step 7* and *step 8* to debug it. You can see a preview here of how the tasks look when an error occurs:

Figure 9.15 – DAG Graph page view showing a running error

See also

You can find the code for the Airflow example DAGs on their official GitHub page here: `https://github.com/apache/airflow/tree/main/airflow/example_dags`.

Creating custom operators

As seen in the previous recipe, *Creating DAGs*, it is nearly impossible to create a DAG without instantiating a task or, in other words, defining an operator. Operators are responsible for holding the logic required to process data in the pipeline.

We also know that Airflow already has predefined operators, allowing dozens of ways to ingest and process data. Now, it is time to put into practice how to create custom operators. Custom operators allow us to apply specific logic to a related project or data pipeline.

You will learn how to create a simple customized operator in this recipe. Although it is very basic, you will be able to apply the foundations of this technique to different scenarios.

In this recipe, we will create a custom operator to connect to and retrieve data from the HolidayAPI, the same as we saw previously, in *Chapter 2*.

Getting ready

Please, refer to the *Getting ready* section in the *Configuring Airflow* recipe for this recipe since we will handle it with the same technology.

We also need to add an environment variable to store our API secret. To do so, select the **Variable** item under the **Admin** menu in the Airflow UI, and you will be redirected to the desired page. Now, click the + button to add a new variable, as follows:

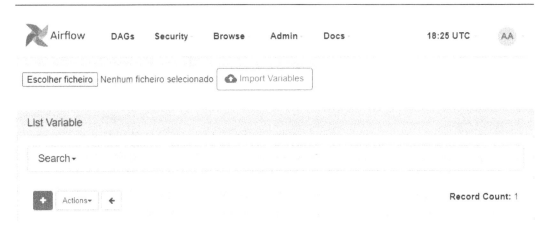

Figure 9.16 – The Variable page in the Airflow UI

On the **List Variable** page, insert SECRET_HOLIDAY_API under the **Key** field and your API secret under the **Value** field. Use the same values you used to execute the *Retrieving data using API authentication* recipe in *Chapter 2*. Save it and you will be redirected to the **Variables** page, as shown in the following screenshot:

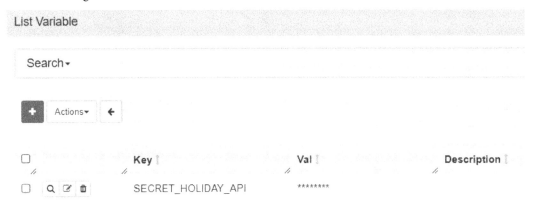

Figure 9.17 – The Airflow UI with a new variable to store the HolidayAPI secret

Now, we are ready to create our custom operator.

How to do it...

The code we will use to create the custom operator is the same one we saw in *Chapter 2*, in the *Retrieving data using API authentication* recipe, with some alterations to fit Airflow's requirements. Here are the steps for it:

1. Let's start by creating the structure inside the `plugins` folder. Since we want to make a custom operator, we need to create a folder called `operators`, where we will put a Python file called `holiday_api_plugin.py`. Your file tree should look like this:

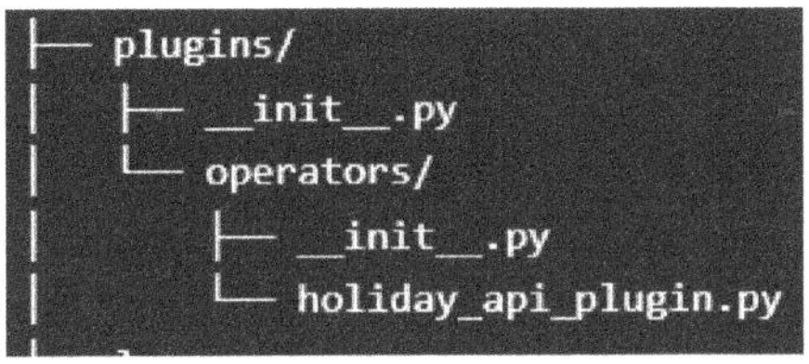

```
├── plugins/
│   ├── __init__.py
│   └── operators/
│       ├── __init__.py
│       └── holiday_api_plugin.py
│
```

Figure 9.18 – Airflow's local directory structure for the plugins folder

2. We will create some code inside `holiday_api_plugin.py`, starting with the library imports and declaring a global variable for where our file output needs to be placed:

```
from airflow.settings import AIRFLOW_HOME
from airflow.models.baseoperator import BaseOperator

import requests
import json

file_output_path = f"{AIRFLOW_HOME}/files_to_test/output_files/"
```

3. Then, we need to create a Python class, declare its constructors, and finally insert the exact code from *Chapter 2* inside a function called `execute`:

```
class HolidayAPIIngestOperator(BaseOperator):
    def __init__(self, filename, secret_key, country, year,
**kwargs):
        super().__init__(**kwargs)
        self.filename = filename
        self.secret_key = secret_key
        self.country = country
        self.year = year

    def execute(self, context):
        params = { 'key': self.secret_key,
                'country': self.country,
```

```
                'year': self.year
        }
        url = "https://holidayapi.com/v1/holidays?"
        output_file = file_output_path + self.filename
        try:
            req = requests.get(url, params=params)
            print(req.json())
            with open(output_file, "w") as f:
                json.dump(req.json(), f)
            return "Holidays downloaded successfully"
        except Exception as e:
            raise e
```

Save the file and our operator is ready. Now, we need to create the DAG to execute it.

4. Using the same logic as in the *Creating DAGs* recipe, we will create a file called `holiday_ingest_dag.py`. Your new DAG directory tree should look like this:

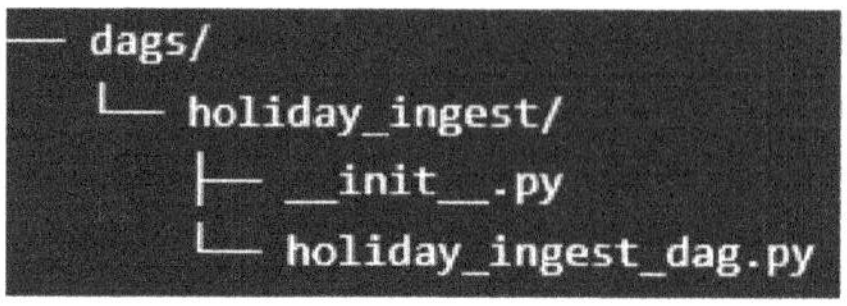

Figure 9.19 – Airflow's directory structure for the holiday_ingest DAG folder

5. Now, let's insert our DAG code inside the `holiday_ingest_dag.py` file and save it:

```
from airflow import DAG
# Other imports
from operators.holiday_api_plugin import
HolidayAPIIngestOperator

# Define default arguments

# Instantiate a DAG object
with DAG(
    dag_id='holiday_ingest',
    default_args=default_args,
    schedule_interval=timedelta(days=1),
) as dag:
    filename_json = f"holiday_brazil.json"
    task = HolidayAPIIngestOperator(
        task_id="holiday_api_ingestion",
        filename=filename_json,
        secret_key=Variable.get("SECRET_HOLIDAY_API"),
```

```
        country="BR",
        year=2022
    )


task
```

For the full code, refer to the GitHub repository here: `https://github.com/PacktPublishing/Data-Ingestion-with-Python-Cookbook/tree/main/Chapter_9/creating_custom_operators`.

6. Then, go to the Airflow UI, look for the `holiday_ingest` DAG, and enable it. It will look like the following figure:

Figure 9.20 – The holiday_ingest DAG enabled in the Airflow UI

Your job will start to run immediately.

7. Now, let's find the task logs by following the same steps from the *Creating DAGs* recipe, but now clicking on the `holiday_api_ingestion` task. Your log page should look like the following figure:

```
*** Reading local file: /opt/airflow/logs/dag_id=holiday_ingest/run_id=manual__2023-03-24T22:49:26.842087+00:00/task_id=holiday_api_ingestion/attempt=1.log
[2023-03-24, 22:49:27 UTC] {taskinstance.py:1159} INFO - Dependencies all met for <TaskInstance: holiday_ingest.holiday_api_ingestion manual__2023-03-24T22:49:26.842087+00:00 [queued]>
[2023-03-24, 22:49:27 UTC] {taskinstance.py:1159} INFO - Dependencies all met for <TaskInstance: holiday_ingest.holiday_api_ingestion manual__2023-03-24T22:49:26.842087+00:00 [queued]>
[2023-03-24, 22:49:27 UTC] {taskinstance.py:1356} INFO -
--------------------------------------------------------------------------------
[2023-03-24, 22:49:27 UTC] {taskinstance.py:1357} INFO - Starting attempt 1 of 2
[2023-03-24, 22:49:27 UTC] {taskinstance.py:1358} INFO -
--------------------------------------------------------------------------------
[2023-03-24, 22:49:27 UTC] {taskinstance.py:1377} INFO - Executing <Task(HolidayAPIIngestOperator): holiday_api_ingestion> on 2023-03-24 22:49:26.842087+00:00
[2023-03-24, 22:49:27 UTC] {standard_task_runner.py:52} INFO - Started process 587 to run task
[2023-03-24, 22:49:27 UTC] {standard_task_runner.py:79} INFO - Running: ['***', 'tasks', 'run', 'holiday_ingest', 'holiday_api_ingestion', 'manual__2023-03-24T22:49:26.842087+00:00', '-
[2023-03-24, 22:49:27 UTC] {standard_task_runner.py:80} INFO - Job 223: Subtask holiday_api_ingestion
[2023-03-24, 22:49:27 UTC] {task_command.py:369} INFO - Running <TaskInstance: holiday_ingest.holiday_api_ingestion manual__2023-03-24T22:49:26.842087+00:00 [running]> on host d3a7d2eef1
[2023-03-24, 22:49:28 UTC] {taskinstance.py:1571} INFO - Exporting the following env vars:
AIRFLOW_CTX_DAG_OWNER=***
AIRFLOW_CTX_DAG_ID=holiday_ingest
AIRFLOW_CTX_TASK_ID=holiday_api_ingestion
AIRFLOW_CTX_EXECUTION_DATE=2023-03-24T22:49:26.842087+00:00
AIRFLOW_CTX_TRY_NUMBER=1
AIRFLOW_CTX_DAG_RUN_ID=manual__2023-03-24T22:49:26.842087+00:00
[2023-03-24, 22:49:28 UTC] {logging_mixin.py:115} INFO - {'status': 200, 'warning': 'These results do not include state and province holidays. For more information, please visit https:/
[2023-03-24, 22:49:28 UTC] {taskinstance.py:1400} INFO - Marking task as SUCCESS. dag_id=holiday_ingest, task_id=holiday_api_ingestion, execution_date=20230324T224928, start_date=202303
[2023-03-24, 22:49:28 UTC] {local_task_job.py:156} INFO - Task exited with return code 0
[2023-03-24, 22:49:28 UTC] {local_task_job.py:273} INFO - 0 downstream tasks scheduled from follow-on schedule check
```

Figure 9.21 – holiday_api_ingestion task logs

8. Finally, let's see whether the output file was created successfully. Go to the `files_to_test` folder, click on the `output_files` folder, and if everything was successfully configured, a file called `holiday_brazil.json` will be inside it. See the following figure for reference:

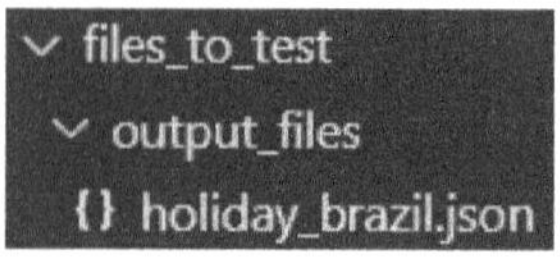

Figure 9.22 – holiday_brazil.json inside the output_files screenshot

The beginning of the output file should look like this:

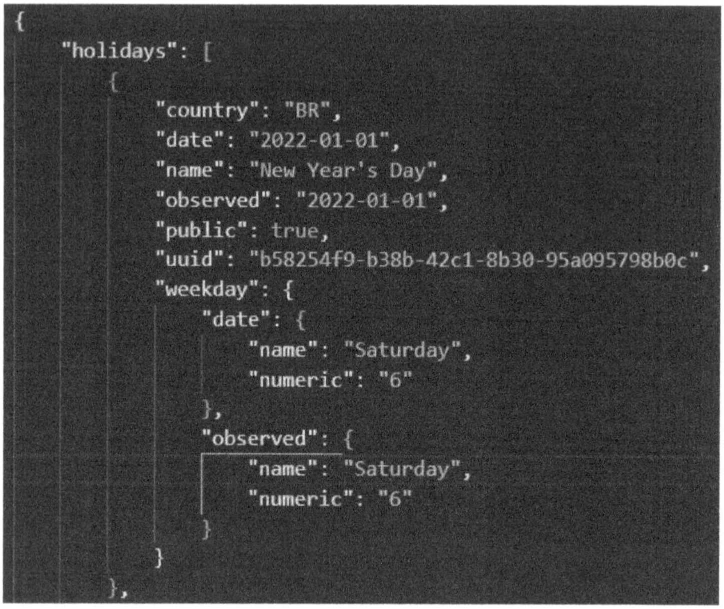

```
{
    "holidays": [
        {
            "country": "BR",
            "date": "2022-01-01",
            "name": "New Year's Day",
            "observed": "2022-01-01",
            "public": true,
            "uuid": "b58254f9-b38b-42c1-8b30-95a095798b0c",
            "weekday": {
                "date": {
                    "name": "Saturday",
                    "numeric": "6"
                },
                "observed": {
                    "name": "Saturday",
                    "numeric": "6"
                }
            }
        },
```

Figure 9.23 – The first lines of holiday_brazil.json

How it works...

As you can see, a custom Airflow operator is an isolated class with a unique purpose. Usually, custom operators are created with the intention to also be used by other teams or DAGs, which avoids code redundancy or duplication. Now, let's understand how it works.

We started the recipe by creating the file to host the new operator inside the `plugin` folder. We do this because, internally, Airflow understands that everything inside of it is custom code. Since we wanted to only create an operator, we put it inside a folder with the same name. However, it is also possible to create another resource called **Hooks**. You can learn more about creating hooks in Airflow here: `https://airflow.apache.org/docs/apache-airflow/stable/howto/custom-operator.html`.

Now, heading to the operator code, we declare our code to ingest the HolidayAPI inside a class, as you can see here:

```
class HolidayAPIIngestOperator(BaseOperator):
    def __init__(self, filename, secret_key, country, year, **kwargs):
        super().__init__(**kwargs)
        self.filename = filename
        self.secret_key = secret_key
        self.country = country
        self.year = year
```

We did this to extend Airflow's `BaseOperator` so that we could customize it and insert new constructors. `filename`, `secret_key`, `country`, and `year` are the parameters we need to execute the API ingest.

Then, we declared the `execute` function to ingest data from the API. The context is an Airflow parameter that allows the function to read configuration values:

```
def execute(self, context):
```

Then, our final step was to create a DAG to execute the operator we made. The code is like the previous DAG we created in the *Creating DAGs* recipe, only with a few new items. The first item was the new `import` instances, as you can see here:

```
from airflow.models import Variable
from operators.holiday_api_plugin import HolidayAPIIngestOperator
```

The first `import` statement allows us to use the value of SECRET_HOLIDAY_API we inserted using the UI, and the second imports our custom operator. Observe that we only used the `operators.holiday_api_plugin` path. Due to Airflow's internal configuration, it understands that the code inside an `operators` folder (inside the `plugins` folder) is an operator.

Now we can instantiate the custom operator like any other built-in operator in Airflow by passing the required parameters, as you can see in the code here:

```
task = HolidayAPIIngestOperator(
        task_id="holiday_api_ingestion",
        filename=filename_json,
        secret_key=Variable.get("SECRET_HOLIDAY_API"),
        country="BR",
        year=2022)
```

There's more...

If an entire project has the same form of authentication for retrieving data from a specific API or database, creating custom operators or hooks is a valuable way to avoid code duplication.

However, before jumping excitedly into creating your plugin, remember that Airflow's community already provides many operators. For example, if you use AWS in your daily work, you don't need to worry about creating a new operator to connect with AWS Glue since that already has been done and approved by the Apache community. See the documentation here: `https://airflow.apache.org/docs/apache-airflow-providers-amazon/stable/operators/glue.html`.

You can see the complete list of AWS operators here: `https://airflow.apache.org/docs/apache-airflow-providers-amazon/stable/operators/index.html`.

See also

For more custom operator examples, see *Virajdatt Kohir's* blog here: `https://kvirajdatt.medium.com/airflow-writing-custom-operators-and-publishing-them-as-a-package-part-2-3f4603899ec2`.

Configuring sensors

Under the operator's umbrella, we have sensors. Sensors are designed to wait to execute a task until something happens. For example, a sensor triggers a pipeline (or task) when a file lands in an HDFS folder, as shown here: `https://airflow.apache.org/docs/apache-airflow-providers-apache-hdfs/stable/_api/airflow/providers/apache/hdfs/sensors/hdfs/index.html`. As you might be wondering, there are also sensors for specific schedules or time deltas.

Sensors are a fundamental part of creating an automated and event-driven pipeline. In this recipe, we will configure a `weekday` sensor, which executes our data pipeline on a specific day of the week.

Getting ready

Refer to the *Getting ready* section in the *Configuring Airflow* recipe for this recipe since we will handle it with the same technology.

Besides that, let's put a JSON file to the following path inside the Airflow folder: `files_to_test/sensors_files/`.

In my case, I will use the `github_events.json` file, but you can use any of your preferences.

How to do it...

Here are the steps to perform this recipe:

1. Let's start our DAG script by importing the required libraries, defining `default_args`, and instantiating our DAG, as you can see here:

```
from airflow import DAG
from airflow.settings import AIRFLOW_HOME
from airflow.operators.bash import BashOperator
from airflow.sensors.weekday import DayOfWeekSensor
from airflow.utils.weekday import WeekDay
from datetime import datetime, timedelta

default_args = {
    'owner': 'airflow',
    'start_date': datetime(2023, 3, 22),
    'retry_delay': timedelta(minutes=5)
}

# Instantiate a DAG object
with DAG(
    dag_id='sensors_move_file',
    default_args=default_args,
    schedule_interval="@once",
) as dag:
```

2. Now, let's define our first task using `DayOfWeekSensor`. See the code here:

```
move_file_on_saturdays = DayOfWeekSensor(
    task_id="move_file_on_saturdays",
    timeout=120,
    soft_fail=True,
    week_day=WeekDay.SATURDAY
)
```

I suggest setting the day of the week as a parameter while doing this exercise to ensure no confusion. For example, if you want it to be executed on a Monday, set `week_day` to `WeekDay.MONDAY`, and so on.

3. Then, we will define another task using `BashOperator`. This task will execute the command to move a JSON file from `files_to_test/sensors_files/` to `files_to_test/output_files/`. Your code should look like this:

```
move_file_task = BashOperator(
        task_id="move_file_task",
```

```
        bash_command="mv $AIRFLOW_HOME/files_to_test/
    sensors_files/*.json $AIRFLOW_HOME/files_to_test/output_files/",
            )
```

4. Then, we will define the execution workflow of our DAG, as you can see here:

```
move_file_on_saturdays.set_downstream(move_file_task)
```

The `.set_downstream()` function will work similarly to the double arrows (`>>`) we already used to define the workflow. You can read more about this here: https://airflow.apache.org/docs/apache-airflow/1.10.3/concepts.html?highlight=trigger#bitshift-composition.

5. As seen in the previous two recipes of this chapter, now we will enable our sensors_move_file DAG, which will start immediately. If you set the weekday as the same day on which you are executing this exercise, your DAG **Graph** view will look like this, indicating success:

Figure 9.24 – sensors_move_file tasks showing a success status

6. Now, let's see whether our file was moved to the directories. As described in the *Getting ready* section, I put a JSON file called github_events.json inside the sensor_files folder. Now, it will be inside output_files, as you can see here:

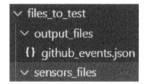

Figure 9.25 – github_events.json inside the output_files folder

This indicates our sensor executed as expected!

How it works...

Sensors are valuable operators that execute an action based on a state. They can be triggered when a file lands in a directory, during the day, when an external task finishes, and so on. Here, we approached an example using a day of the week commonly used in data teams to change files from an ingested folder to a cold storage folder.

Sensors count with an internal method called `poke`, which will check a resource's status until the criteria are met. If you look at the `move_file_on_saturday` log, you will see something like this:

```
[2023-03-25, 00:19:03 UTC] {weekday.py:83} INFO - Poking until weekday
is in WeekDay.SATURDAY, Today is SATURDAY
[2023-03-25, 00:19:03 UTC] {base.py:301} INFO - Success criteria met.
Exiting.
[2023-03-25, 00:19:03 UTC] {taskinstance.py:1400} INFO - Marking task
as SUCCESS. dag_id=sensors_move_file, task_id=move_file_on_saturdays,
execution_date=20230324T234623, start_date=20230325T001903, end_
date=20230325T001903
[2023-03-25, 00:19:03 UTC] {local_task_job.py:156} INFO - Task exited
with return code 0
```

Looking at the following code, we did not define a `reschedule` parameter, so the job will stop until we manually trigger it again:

```
move_file_on_saturdays = DayOfWeekSensor(
    task_id="move_file_on_saturdays",
    timeout=120,
    soft_fail=True,
    week_day=WeekDay.SATURDAY
)
```

Other parameters we defined were `timeout`, which indicates the time in seconds before it fails or stops retrying, and `soft_fail`, which marks the task as `SKIPPED` in the case of failure.

You can see other allowed parameters here: `https://airflow.apache.org/docs/apache-airflow/stable/_api/airflow/sensors/base/index.html`.

And, of course, like the rest of the operators, we can create our custom sensor by extending the `BaseSensorOperator` class from Airflow. The main challenge here is that to be considered a sensor, it needs to overwrite the `poke` parameter without creating a recursing or non-ending function.

See also

You can see a list of the default Airflow sensors on the official documentation page here: `https://airflow.apache.org/docs/apache-airflow/stable/_api/airflow/sensors/index.html`.

Creating connectors in Airflow

Having DAGs and operators without connecting to any external source is useless. Of course, there are many ways to ingest files, even from other DAGs or task results. Still, data ingestion usually involves using external sources such as APIs or databases as the first step of a data pipeline.

To make this happen, in this recipe, we will understand how to create a connector in Airflow to connect to a sample database.

Getting ready

Refer to the *Getting ready* section of the *Configuring Airflow* recipe for this recipe since we will handle it with the same technology.

This exercise will also require the MongoDB local database to be up and running. Ensure you have configured it as seen in *Chapter 1* and have at least one database and collection. You can use the instructions from the *Connecting to a NoSQL database (MongoDB)* recipe in *Chapter 5*.

How to do it...

Here are the steps to perform this recipe:

1. Let's start by opening the Airflow UI. On the top menu, select the **Admin** button and then **Connections**, and you will be redirected to the **Connections** page. Since we haven't configured anything yet, this page will be empty, as you can see in the following screenshot:

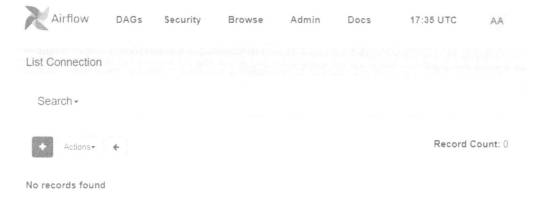

Figure 9.26 – The Connections page in the Airflow UI

2. Then, click the + button to be redirected to the **Add Connection** page. Under the **Connection Type** field, search for and select **MongoDB**. Insert your connection values under the respective fields, as shown in the following screenshot:

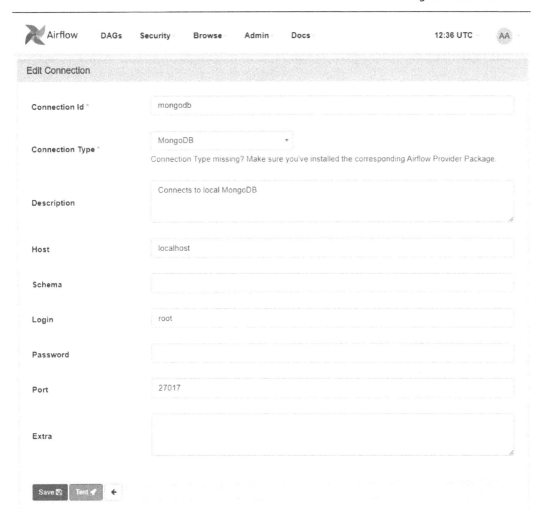

Figure 9.27 – Creating a new connector in the Airflow UI

Click on the **Save** button, and you should have something similar to this on the **Connection** page:

		Conn Id	Conn Type	Description	Host	Port	Is Encrypted	Is Extra Encrypted
☐	☑ 🗑	mongodb	mongo	Connects to local MongoDB	localhost	27017	False	False

Figure 9.28 – The MongoDB connector created in the Airflow UI

3. Let's create our new DAG using the same folder and file tree structure that we saw in the *Creating DAGs* recipe. I will call the `mongodb_check_conn_dag.py` DAG file.

4. Inside the DAG file, let's start by importing and declaring the required libraries and variables, as you can see here:

```
from airflow import DAG
from airflow.settings import AIRFLOW_HOME
from airflow.providers.mongo.hooks.mongo import MongoHook
from airflow.operators.python import PythonOperator

import os
import json
from datetime import datetime, timedelta

default_args = {
    'owner': 'airflow',
    'depends_on_past': False,
    'start_date': datetime(2023, 3, 22),
    'retries': 1,
    'retry_delay': timedelta(minutes=5)
}
```

5. Now, we will create a function to connect with MongoDB locally and print `collection` `reviews` from the db_airbnb database, as you can see here:

```
def get_mongo_collection():
    hook = MongoHook(conn_id ='mongodb')
    client = hook.get_conn()
    print(client)
    print( hook.get_collection(mongo_collection="reviews",
mongo_db="db_airbnb"))
```

6. Then, let's proceed with the DAG object:

```
# Instantiate a DAG object
with DAG(
    dag_id='mongodb_check_conn',
    default_args=default_args,
    schedule_interval=timedelta(days=1),
) as dag:
```

7. Finally, let's use `PythonOperator` to call our `get_mongo_collection` function defined in *step 5*:

```
mongo_task = PythonOperator(
    task_id='mongo_task',
    python_callable=get_mongo_collection
)
```

Don't forget to put the name of your task in the indentation of the DAG, as follows:

```
mongo_task
```

8. Heading to the Airflow UI, let's enable the DAG and wait for it to be executed. After finishing successfully, your `mongodb_task` log should look like this:

```
*** Reading local file: /opt/airflow/logs/dag_id=mongodb_check_conn/run_id=scheduled__2023-03-24T00:00:00+00:00/task_id=mongo_task/attempt=1.log
[2023-03-25, 12:37:16 UTC] {taskinstance.py:1159} INFO - Dependencies all met for <TaskInstance: mongodb_check_conn.mongo_task scheduled__2023-03-24T00:00:00+00:00 [queued]>
[2023-03-25, 12:37:16 UTC] {taskinstance.py:1159} INFO - Dependencies all met for <TaskInstance: mongodb_check_conn.mongo_task scheduled__2023-03-24T00:00:00+00:00 [queued]>
[2023-03-25, 12:37:16 UTC] {taskinstance.py:1356} INFO -
--------------------------------------------------------------------------------
[2023-03-25, 12:37:16 UTC] {taskinstance.py:1357} INFO - Starting attempt 1 of 2
[2023-03-25, 12:37:16 UTC] {taskinstance.py:1358} INFO -
--------------------------------------------------------------------------------
[2023-03-25, 12:37:16 UTC] {taskinstance.py:1377} INFO - Executing <Task(PythonOperator): mongo_task> on 2023-03-24 00:00:00+00:00
[2023-03-25, 12:37:16 UTC] {standard_task_runner.py:52} INFO - Started process 3158 to run task
[2023-03-25, 12:37:16 UTC] {standard_task_runner.py:79} INFO - Running: ['***', 'tasks', 'run', 'mongodb_check_conn', 'mongo_task', 'scheduled__2023-03-24T00:00:00+00:00', '--job-id
[2023-03-25, 12:37:16 UTC] {standard_task_runner.py:80} INFO - Job 267: Subtask mongo_task
[2023-03-25, 12:37:16 UTC] {task_command.py:369} INFO - Running <TaskInstance: mongodb_check_conn.mongo_task scheduled__2023-03-24T00:00:00+00:00 [running]> on host a7644993644c
[2023-03-25, 12:37:16 UTC] {taskinstance.py:1571} INFO - Exporting the following env vars:
AIRFLOW_CTX_DAG_OWNER=***
AIRFLOW_CTX_DAG_ID=mongodb_check_conn
AIRFLOW_CTX_TASK_ID=mongo_task
AIRFLOW_CTX_EXECUTION_DATE=2023-03-24T00:00:00+00:00
AIRFLOW_CTX_TRY_NUMBER=1
AIRFLOW_CTX_DAG_RUN_ID=scheduled__2023-03-24T00:00:00+00:00
[2023-03-25, 12:37:16 UTC] {base.py:68} INFO - Using connection ID 'mongodb' for task execution.
[2023-03-25, 12:37:16 UTC] {logging_mixin.py:115} INFO - MongoClient(host=['localhost:27017'], document_class=dict, tz_aware=False, connect=True)
[2023-03-25, 12:37:16 UTC] {logging_mixin.py:115} INFO - Collection(Database(MongoClient(host=['localhost:27017'], document_class=dict, tz_aware=False, connect=True), 'db_airbnb'),
[2023-03-25, 12:37:16 UTC] {python.py:173} INFO - Done. Returned value was: None
[2023-03-25, 12:37:16 UTC] {taskinstance.py:1400} INFO - Marking task as SUCCESS. dag_id=mongodb_check_conn, task_id=mongo_task, execution_date=20230324T000000, start_date=20230325T
[2023-03-25, 12:37:16 UTC] {local_task_job.py:156} INFO - Task exited with return code 0
[2023-03-25, 12:37:16 UTC] {local_task_job.py:273} INFO - 0 downstream tasks scheduled from follow-on schedule check
```

Figure 9.29 – mongodb_task logs

As you can see, we connected and retrieved the `Collection` object from MongoDB.

How it works...

Creating a connection in Airflow is straightforward, as demonstrated here using the UI. It is also possible to create connections programmatically using the `Connection` class.

After we set our MongoDB connection parameters, we needed to create a form to access it, and we did so using a hook. A **hook** is a high-level interface that allows connections to external sources without the need to be preoccupied with low-level code or special libraries.

Remember that we configured an external package in the *Configuring Airflow* recipe? It was a provider to allow an easy connection with MongoDB:

```
from airflow.providers.mongo.hooks.mongo import MongoHook
```

Inside the `get_mongo_collection` function, we instantiated `MongoHook` and passed the same connection ID name set in the `Connection` page, as you can see here:

```
hook = MongoHook(conn_id ='mongodb')
```

With that instance, we can call the methods of the `MongoHook` class and even pass other parameters to configure the connection. See the documentation for this class here: https://airflow.apache.org/docs/apache-airflow-providers-mongo/stable/_api/airflow/providers/mongo/hooks/mongo/index.html.

There's more...

You can also use the `airflow.cfg` file to set the connection strings or any other environment variable. It is a good practice to store sensitive information here, such as credentials, since they will not be shown in the UI. It is also possible to encrypt these values with additional configuration.

For more information, see the documentation here: `https://airflow.apache.org/docs/apache-airflow/stable/howto/set-config.html`.

See also

- You can learn about the MongoDB provider on the Airflow official documentation page here: `https://airflow.apache.org/docs/apache-airflow-providers-mongo/stable/_api/airflow/providers/mongo/index.html`

- If you are interested in reading more about connections, see this link: `https://airflow.apache.org/docs/apache-airflow/stable/howto/connection.html#custom-connection-types`

Creating parallel ingest tasks

When working with data, we hardly ever just perform one ingestion at a time, and a real-world project involves many ingestions happening simultaneously, often in parallel. We know scheduling two or more DAGs to run alongside each other is possible, but what about tasks inside one DAG?

This recipe will illustrate how to create parallel task execution in Airflow.

Getting ready

Please refer to the *Getting ready* section of the *Configuring Airflow* recipe for this recipe since we will handle it with the same technology.

To avoid redundancy in this exercise, we won't explicitly include the imports and main DAG configuration. Instead, the focus is on organizing the operator's workflow. You can use the same logic to create your DAG structure as in the *Creating DAGs* recipe.

For the complete Python file used here, go to the GitHub page here: `https://github.com/PacktPublishing/Data-Ingestion-with-Python-Cookbook/tree/main/Chapter_9/creating_parallel_ingest_tasks`.

How to do it...

For this exercise, my DAG will be called `parallel_tasks_dag`. Now, let's try it:

1. Let's start by creating five `BashOperator` instances, as you can see here:

```
# (DAG configuration above)
    t_0 = BashOperator(
            task_id="t_0",
            bash_command="echo 'This tasks will be executed
first'",
        )

    t_1 = BashOperator(
            task_id="t_1",
            bash_command="echo 'This tasks no1 will be executed
in parallel with t_2 and t_3'",
        )

    t_2 = BashOperator(
            task_id="t_2",
            bash_command="echo 'This tasks no2 will be executed
in parallel with t_1 and t_3'",
        )

    t_3 = BashOperator(
            task_id="t_3",
            bash_command="echo 'This tasks no3 will be executed
in parallel with t_1 and t_2'",
        )

    t_final = BashOperator(
        task_id="t_final",
        bash_command="echo 'Finished all tasks in parallel'",
    )
```

2. The idea is for three of them to be executed in parallel so they will be inside square brackets. The first and last tasks will have the same workflow declared as we saw in the *Creating DAGs* recipe, using the `>>` character. The final flow structure will look like this:

```
t_0 >> [t_1, t_2, t_3] >> t_final
```

3. Finally, enable your DAG, and let's see what it looks like on the DAG **Graph** page. It would be best if you had something like the following figure:

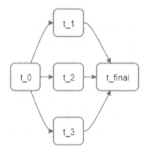

Figure 9.30 – parallel_tasks_dag tasks in Airflow

As you can observe, the tasks inside the square brackets are displayed in parallel and will start after t_0 finishes its work.

How it works...

Although creating parallel tasks inside a DAG is simple, this type of workflow arrangement is advantageous when working with data.

Consider an example of data ingestion: we need to guarantee we have ingested all the desired endpoints before moving out to the next pipeline phase. See the following figure as a reference:

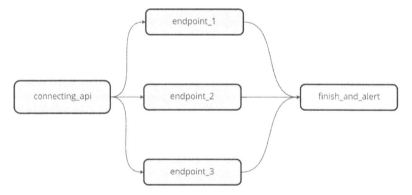

Figure 9.31 – Example of parallel execution

The parallel execution will only move to the final task when all the parallel ones finish successfully. With this, we guarantee the data pipeline will not ingest only a small portion of the data but all the required data.

Back to our exercise, we can simulate this behavior, creating an **End-of-Line** (**EOL**) error in t_2 by removing one of the simple quotation marks. In the following figure, you can see what the DAG graph will look like:

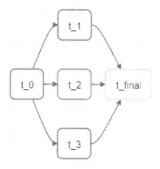

Figure 9.32 – Airflow's parallel_tasks_dag with an error t_2 task

The t_final task will retry executing until we fix t_2 or the number of retries reaches its limit.

However, avoiding many parallel tasks is a good practice, mainly if you have limited infrastructure resources to handle them. There are many ways to create dependency on external tasks or DAGs, and we can use them to make more efficient pipelines.

There's more...

Along with the concept of task parallelism, we have BranchOperator. BranchOperator executes one or more tasks simultaneously based on a criteria match. Let's illustrate this with the following figure:

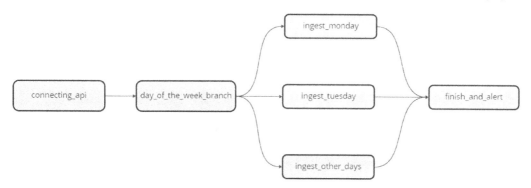

Figure 9.33 – Branching task diagram example

Based on the day-of-the-week criteria, the day_of_the_week_branch task will trigger a specific task assigned for that day.

If you want to know more about it, *Analytics Vidhya* has a good blog post about it, which you can read here: `https://www.analyticsvidhya.com/blog/2023/01/data-engineering-101-branchpythonoperator-in-apache-airflow/`.

See also

- BetterDataScience has a good blog post about parallel tasks in Airflow. You can find it here: `https://betterdatascience.com/apache-airflow-run-tasks-in-parallel/`.

- You can read more about Airflow task parallelism here: `https://hevodata.com/learn/airflow-parallelism/`.

Defining ingest-dependent DAGs

In the data world, considerable discussion exists about how to organize Airflow DAGs. The approach I generally use is to create a DAG for a specific pipeline based on the business logic or final destination. Nevertheless, sometimes, to proceed with a task inside a DAG, we depend on another DAG to finish the process and get the output.

In this recipe, we will create two DAGs, where the first depends on the result of the second to be successful. Otherwise, it will not be completed. To assist us, we will use the `ExternalTaskSensor` operator.

Getting ready

Please refer to the *Getting ready* section of the *Configuring Airflow* recipe for this recipe since we will handle it with the same technology.

This recipe depends on the `holiday_ingest` DAG, created in the *Creating custom operators* recipe, so ensure you have that.

We will not explicitly cite the imports and main DAG configuration to prevent redundancy and repetition in this exercise. The aim here is how to organize the operator's workflow. You can use the same logic to create your DAG structure as in the *Creating DAGs* recipe.

For the complete Python file used here, go to the GitHub page here:

`https://github.com/PacktPublishing/Data-Ingestion-with-Python-Cookbook/tree/main/Chapter_9/de%EF%AC%81ning_dependent_ingests_DAGs`

How to do it...

For this exercise, let's create a DAG triggered when `holiday_ingest` finishes successfully, and returns all the holiday dates in the console output. My DAG will be called `external_sensor_dag`, but feel free to provide any other ID name. Just ensure it is unique and therefore will not impair other DAGs:

1. Along with the default imports we have, let's add the following:

    ```
    from airflow.sensors.external_task import ExternalTaskSensor
    ```

2. Now, we will insert a Python function to return the holiday dates in the `holiday_brazil.json` file, which is the output of the `holiday_ingest` DAG:

    ```
    def get_holiday_dates(filename_json):
        with open (filename_json, 'r') as f:
            json_hol = json.load(f)
            holidays = json_hol["holidays"]
        print([item['date'] for item in holidays])
    ```

3. Then, we will make the two operators of the DAG and define the workflow:

    ```
    wait_holiday_api_ingest = ExternalTaskSensor(
        task_id='wait_holiday_api_ingest',
        external_dag_id='holiday_ingest',
        external_task_id='holiday_api_ingestion',
        allowed_states=["success"],
        execution_delta = timedelta(minutes=1),
        timeout=600,
    )


    filename_json = f"{AIRFLOW_HOME}/files_to_test/output_files/
    holiday_brazil.json"
    date_tasks = PythonOperator(
        task_id='date_tasks',
        python_callable=get_holiday_dates,
        op_args=[filename_json]
    )


    wait_holiday_api_ingest >> date_tasks
    ```

Save it and enable this DAG in the Airflow UI. Once enabled, you will notice the `wait_holiday_api_ingest` task will be in the `RUNNING` state and will not proceed to the other task, as follows:

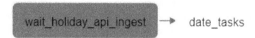

Figure 9.34 – The wait_holiday_api_ingest task in the running state

You will also notice the log for this task looks like the following:

```
[2023-03-26, 20:50:23 UTC] {external_task.py:166} INFO - Poking
for tasks ['holiday_api_ingestion'] in dag holiday_ingest on
2023-03-24T23:50:00+00:00 ...
[2023-03-26, 20:51:23 UTC] {external_task.py:166} INFO - Poking
for tasks ['holiday_api_ingestion'] in dag holiday_ingest on
2023-03-24T23:50:00+00:00 ...
```

4. Now, we will enable and run `holiday_ingest` (if it is not enabled yet).

5. Then, go back to `external_sensor_dag`, and you will see it finished successfully, as shown in the following figure:

Figure 9.35 – external_sensor_dag showing success

If we examine the logs of `date_tasks`, you will see the following output on the console:

```
[2023-03-25, 16:39:31 UTC] {logging_mixin.py:115} INFO - ['2022-
01-01', '2022-02-28', '2022-03-01', '2022-03-02', '2022-03-20',
'2022-04-15', '2022-04-17', '2022-04-21', '2022-05-01', '2022-
05-08', '2022-06-16', '2022-06-21', '2022-08-14', '2022-09-07',
'2022-09-23', '2022-10-12', '2022-11-02', '2022-11-15', '2022-
12-21', '2022-12-24', '2022-12-25', '2022-12-31']
```

Here is the complete log for reference:

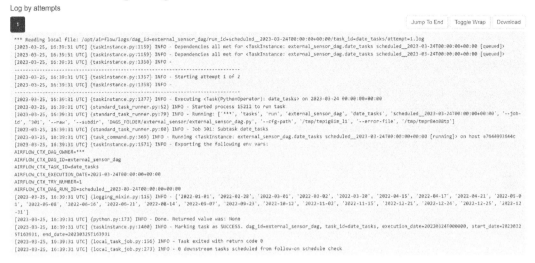

Figure 9.36 – date_tasks logs in the Airflow UI

Now, let's understand how it works in the next section.

How it works...

Let's start by taking a look at our wait_holiday_api_ingest task:

```
wait_holiday_api_ingest = ExternalTaskSensor(
        task_id='wait_holiday_api_ingest',
        external_dag_id='holiday_ingest',
        external_task_id='holiday_api_ingestion',
        allowed_states=["success"],
        execution_delta = timedelta(minutes=1),
        timeout=300,
    )
```

ExternalTaskSensor is a sensor type that will only execute if another task outside its DAG finishes with a specific status defined on the allowed_states parameter. The default value for this parameter is SUCCESS.

The sensor will search for a specific DAG and task in Airflow using the external_dag_id and external_task_id parameters, which we have defined as holiday_ingest and holiday_api_ingestion, respectively. Finally, execution_delta will determine the time interval at which to poke the external DAG again.

Once it finishes, the DAG will remain in the SUCCESS state unless we define a different behavior in the default arguments. If we clear its status, it will return to the RUNNING mode until the sensor criteria are met again.

There's more...

We know Airflow is a powerful tool, but it is not immune from occasional failures. Internally, Airflow has its routes to reach internal and external DAGs, which can occasionally fail. For example, one of these errors might be a DAG not being found, which can happen due to various reasons such as misconfiguration or connectivity issues. You can see a screenshot of one of these errors here:

Figure 9.37 – Occasional 403 error log in an Airflow task

Looking closely, we can observe that for several seconds, one of the Airflow workers lost permission to access or retrieve information from another worker. If this happens to you, disable your DAG and enable it again.

XCom

In this exercise, we used an output file to perform an action, but we can also use the output of a task without requiring it to write a file somewhere. Instead, we can use the **XCom** (short for **cross-communications**) mechanism to help us with it.

To use XCom across tasks, we can simply use *push* and *pull* the values using the xcom_push and xcom_pull methods inside the required tasks. Behind the scenes, Airflow stores those values temporarily in one of its databases, making it easier to access them again.

To check your stored XComs in the Airflow UI, click on the **Admin** button and select **XCom**.

> **Note**
> In production environments, XComs might have a purge routine. Check with the Airflows administrators if you need to keep a value for longer.

You can read more about XComs on the official documentation page here: https://airflow.apache.org/docs/apache-airflow/stable/core-concepts/xcoms.html.

See also

You can learn more about this operator on the Airflow official documentation page: `https://airflow.apache.org/docs/apache-airflow/stable/howto/operator/external_task_sensor.html`.

Further reading

- `https://airflow.apache.org/docs/apache-airflow/stable/core-concepts/dags.html`

- `https://airflow.apache.org/docs/apache-airflow/2.2.4/best-practices.html`

- `https://www.qubole.com/tech-blog/apache-airflow-tutorial-dags-tasks-operators-sensors-hooks-xcom`

- `https://python.plainenglish.io/apache-airflow-how-to-correctly-setup-custom-plugins-2f80fe5e3dbe`

- `https://copyprogramming.com/howto/airflow-how-to-mount-airflow-cfg-in-docker-container`

10

Logging and Monitoring Your Data Ingest in Airflow

We already know how vital logging and monitoring are to manage applications and systems, and Airflow is no different. In fact, **Apache Airflow** already has built-in modules to create logs and export them. But what about improving them?

In the previous chapter, *Putting Everything Together with Airflow,* we covered the fundamental aspects of Airflow, how to start our data ingestion, and how to orchestrate a pipeline and use the best data development practices. Now, let's put into practice the best techniques to enhance logging and monitor Airflow pipelines.

In this chapter, you will learn the following recipes:

- Creating basic logs in Airflow
- Storing log files in a remote location
- Configuring logs in `airflow.cfg`
- Designing advanced monitoring
- Using notification operators
- Using SQL operators for data quality

Technical requirements

You can find the code for this chapter in the GitHub repository here: `https://github.com/PacktPublishing/Data-Ingestion-with-Python-Cookbook`.

Installing and running Airflow

This chapter requires that Airflow is installed on your local machine. You can install it directly on your **operating system** (**OS**) or by using a Docker image. For more information, refer to the *Configuring Docker for Airflow* recipe in *Chapter 1*.

After following the steps described in *Chapter 1*, ensure your Airflow runs correctly. You can do that by checking the Airflow UI here: `http://localhost:8080`.

If you are using a Docker container (as I am) to host your Airflow application, you can check its status on the terminal by running the following command:

```
$ docker ps
```

You can see the command running here:

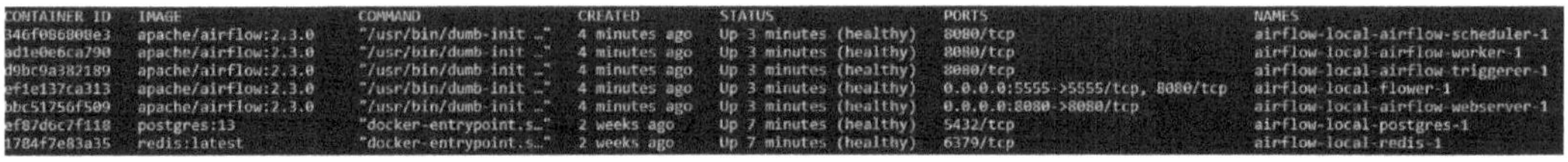

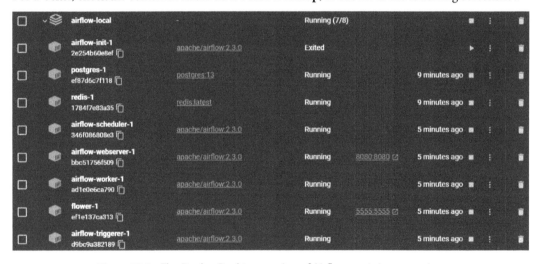

Figure 10.1 – Airflow containers running

For Docker, check the container status on **Docker Desktop**, as shown in the following screenshot:

Figure 10.2 – The Docker Desktop version of Airflow containers running

Airflow environment variables in docker-compose

This section is aimed at users with Airflow running in a Docker container. If you install it directly on your machine, you can skip it.

We need to configure or change Airflow environment variables to complete most of the recipes in this chapter. This kind of configuration is supposed to be done by editing the `airflow.cfg` file. However, this can be tricky if you opt to run your Airflow application using `docker-compose`.

Ideally, we should be able to access the `airflow.cfg` file by mounting a volume in `docker-compose.yaml`, as follows:

```
volumes:
  - ./dags:/opt/airflow/dags
  - ./logs:/opt/airflow/logs
  - ./plugins:/opt/airflow/plugins
  - ./files_to_test:/opt/airflow/files_to_test
  - ./config/airflow.cfg:/opt/airflow/airflow.cfg
```

Figure 10.3 – docker-compose.yaml volumes

Nevertheless, instead of reflecting the file in the local machine, it creates a directory named `airflow.cfg`. It is a bug known by the community (see `https://github.com/puckel/docker-airflow/issues/571`) with no resolution.

To work around it, we will set all the `airflow.cfg` configurations in `docker-compose.yaml` using the environment variables, as shown in the following example:

```
# Remote logging configuration
AIRFLOW__LOGGING__REMOTE_LOGGING: "True"
```

For users who install and run Airflow directly on their local machine, you can proceed by following the steps that instruct you how to edit the `airflow.cfg` file.

Creating basic logs in Airflow

The internal Airflow logging library is based on the Python built-in logs, which provide flexible and configurable forms to capture and store log messages using different components of **directed acyclic graphs** (**DAGs**). Let's start this chapter by covering the basic concepts of how Airflow logs work. This knowledge will allow us to apply more advanced concepts and create mature data ingestion pipelines in real-life projects.

In this recipe, we will create a simple DAG to generate logs based on the default configurations of Airflow. We will also understand how Airflow internally sets the logging architecture.

Getting ready

Refer to the *Technical requirements* section for this recipe, since we will handle it with the same technology.

Since we will create a new DAG, let's create a folder under the dag/ directory called basic_logging and a file inside it called basic_logging_dag.py to insert our script. By the end, your folder structure should look like the following:

```
└── your_airflow_folder/
    ├── dags/
    │   ├── __init__.py
    │   └── basic_logging/
    │       ├── __init__.py
    │       └── basic_logging_dag.py
    ├── plugins
    ├── logs
    ├── files_to_test
    ├── .env
    └── docker-compose.yaml
```

Figure 10.4 – An Airflow directory with a basic_logging DAG structure

How to do it...

The goal is to understand how to create logs in Airflow properly so that the DAG script will be pretty straightforward:

1. Let's start by importing the Airflow and Python libraries:

```
from airflow import DAG
from airflow.operators.python_operator import PythonOperator
from datetime import datetime, timedelta
import logging
```

2. Then, let's get the log configuration we want to use:

```
# Defining the log configuration
logger = logging.getLogger("airflow.task")
```

3. Now, let's define `default_args` and the DAG object which Airflow can create:

```
default_args = {
    'owner': 'airflow',
    'depends_on_past': False,
    'start_date': datetime(2023, 4, 1),
    'retries': 1,
    'retry_delay': timedelta(minutes=5)
}

dag = DAG(
    'basic_logging_dag',
    default_args=default_args,
    description='A simple ETL job using Python and Airflow',
    schedule_interval=timedelta(days=1),
)
```

> **Note**
>
> Unlike in *Chapter 9*, here we will define which tasks belong to this DAG by assigning them to the operator instantiation in *step 5* of this recipe.

4. Now, let's create three example functions only to return log messages. The functions will be named after the ETL steps, as you can see here:

```
def extract_data():
    logger.info("Let's extract data")
    pass

def transform_data():
    logger.info("Then transform data")
    pass

def load_data():
    logger.info("Finally load data")
    logger.error("Oh, where is the data?")
    pass
```

Feel free to insert more log levels if you want to.

5. For each function, we will set a task using `PythonOperator` and the execution order:

```
extract_task = PythonOperator(
    task_id='extract_data',
    python_callable=extract_data,
    dag=dag,
```

```
    )

    transform_task = PythonOperator(
        task_id='transform_data',
        python_callable=transform_data,
        dag=dag,
    )

    load_task = PythonOperator(
        task_id='load_data',
        python_callable=load_data,
        dag=dag,
    )

    extract_task >> transform_task >> load_task
```

You can see that we referred the DAG to each task by assigning the dag object (defined in *step 4*) to a dag parameter.

Save the file and go to the Airflow UI.

6. In the Airflow UI, look for the **basic_logging_dag** DAG and enable it by clicking the toggle button. The job will start right away, and if you check the **Graph** vision of the DAG, you should see something similar to the following screenshot:

Figure 10.5 – The DAG Graph view showing the successful state of the tasks

It means the pipeline ran successfully!

7. Let's check the logs/ directory on our local machine. This directory is at the same level as the DAGs folder, where we put our scripts.

8. You can see more folders inside if you open the `logs/` folder. Look for the one beginning with `dag_id= basic_logging` and open it.

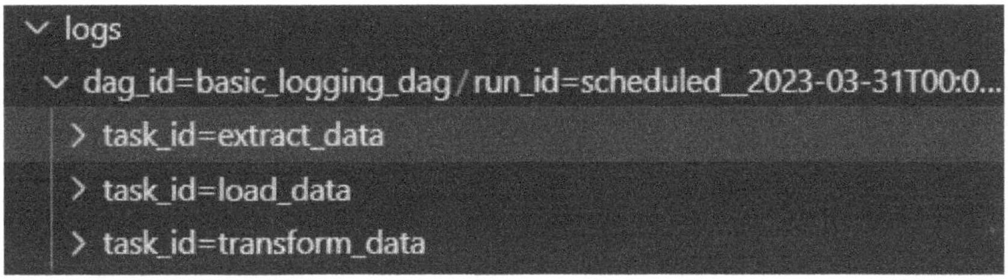

Figure 10.6 – The Airflow logs folder for the basic_logging DAG and its tasks

9. Now, select the folder named `task_id=transform_data` and open the log file inside. You should see something like the following screenshot:

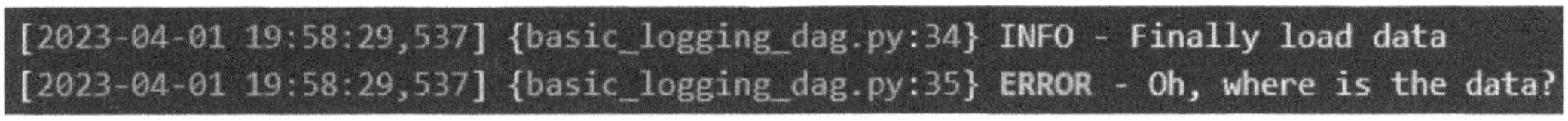

Figure 10.7 – Log messages for the transform_data task

As you can see, the logs were printed on the output and even colored accordingly with the log level, where **INFO** is in green and **ERROR** is in red.

How it works...

This exercise was straightforward, but what if I told you that many developers struggle to understand how Airflow creates its logs? It often happens for two reasons – developers are used to inserting `print()` functions instead of logging methods and only check the records in the Airflow UI.

Depending on the Airflow configuration, it will not show `print()` messages on the UI, and messages used to debug or find where the code ran can be lost. Also, the Airflow UI has a limit on the number of record lines to show, and Spark error messages can be easily omitted in this case.

That's why it is vital to understand that, by default, Airflow stores all its logs under a `logs/` directory, even organizing it by `dag_id`, `run_id`, and each task separately, as we saw in *step 7*. This folder structure can also be changed or improved depending on your needs, and all you need to do is alter the `log_filename_template` variable in `airflow.cfg`. The following is how it is set by default:

```
# Formatting for how airflow generates file names/paths for each task
run.
log_filename_template = dag_id={{ ti.dag_id }}/run_id={{ ti.run_id
}}/task_id={{ ti.task_id }}/{%% if ti.map_index >= 0 %%}map_index={{
ti.map_index }}/{%% endif %%}attempt={{ try_number }}.log
```

Now, looking inside the log file, you can see that it is the same as what is on the UI, as shown in the following screenshot:

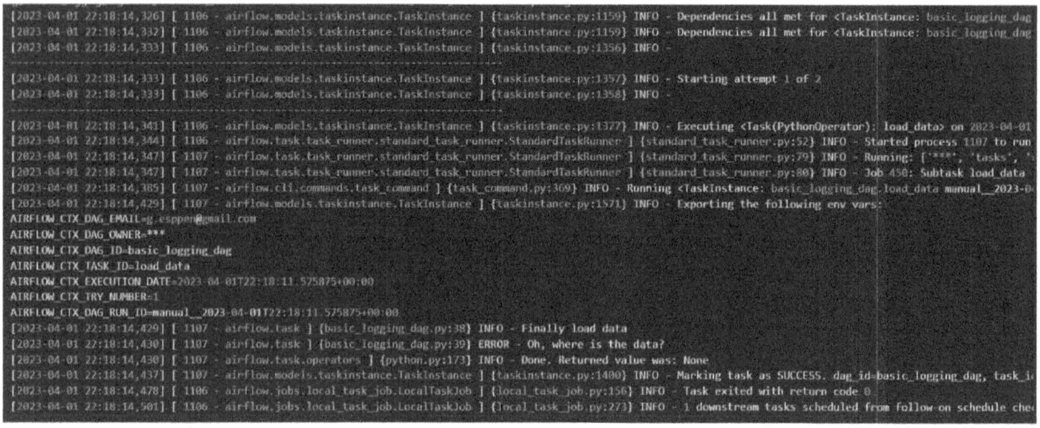

Figure 10.8 – A complete log message stored in a log file found in the local Airflow log folder

In the first lines, it is possible to see the internal calls Airflow makes to start a task, and even the specific function names, such as `taskinstance.py` or `standard_task_runner.py`. Those are all internal scripts. Then, we can see our log messages below in the file.

If you look closely, you can see that the format for our logs is similar to the Airflow core. It happens for two reasons:

- At the beginning of our code, we used the `getLogger()` method to retrieve the configuration used by the `airflow.task` module, as you can see here:

    ```
    logger = logging.getLogger("airflow.task")
    ```

- `airflow.task` uses the Airflow default configuration to format all logs, which can also be found inside the `airflow.cfg` file. Don't worry about this now; we will cover it later in the *Configuring logs in airflow.cfg* recipe.

After defining the `logger` variable and setting the logging class configurations, the rest of the script is straightforward.

See also

You can read more details about Airflow logs on the Astronomer page here: `https://docs.astronomer.io/learn/logging`.

Storing log files in a remote location

By default, Airflow stores and organizes its logs in a local folder with easy access for developers, which facilitates the debugging process when something does not go as expected. However, working with larger projects or teams makes giving everyone access to an Airflow instance or server almost impracticable. Besides looking at the DAG console output, there are other ways to allow access to the logging folder without granting access to Airflow's server.

One of the most straightforward solutions is to export logs to external storage, such as S3 or **Google Cloud Storage**. The good news is that Airflow already has native support to export records to cloud resources.

In this recipe, we will set a configuration in our `airflow.cfg` file that allows the use of the remote logging feature and test it using an example DAG.

Getting ready

Refer to the *Technical requirements* section for this recipe.

AWS S3

To complete this exercise, it is necessary to create an **AWS S3** bucket. Here are the steps required to accomplish it:

1. Create an AWS account by following the steps here: `https://docs.aws.amazon.com/accounts/latest/reference/manage-acct-creating.html`

2. Then, proceed to create an S3 bucket, guided by the AWS documentation here: `https://docs.aws.amazon.com/AmazonS3/latest/userguide/creating-bucket.html`

In my case, I will create an S3 bucket called `airflow-cookbook` for use in this recipe, as you can see in the following screenshot:

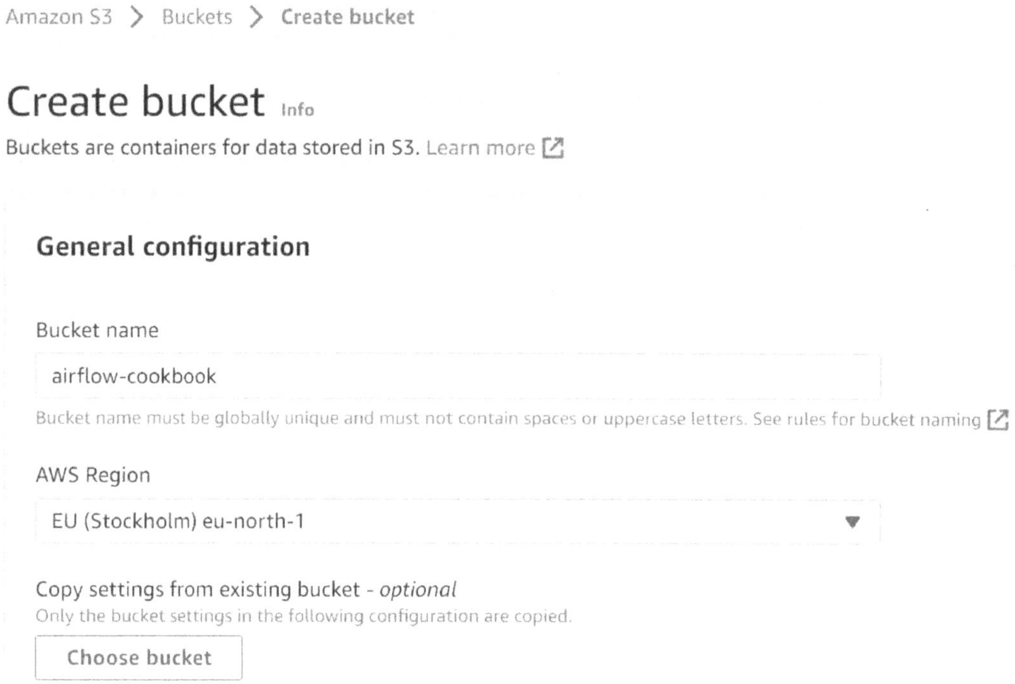

Amazon S3 > Buckets > **Create bucket**

Create bucket Info

Buckets are containers for data stored in S3. Learn more ☑

General configuration

Bucket name

| airflow-cookbook |

Bucket name must be globally unique and must not contain spaces or uppercase letters. See rules for bucket naming ☑

AWS Region

| EU (Stockholm) eu-north-1 ▼ |

Copy settings from existing bucket - *optional*
Only the bucket settings in the following configuration are copied.

| Choose bucket |

Figure 10.9 – The AWS S3 Create bucket page

Airflow DAG code

To avoid redundancy and focus on the goal of this recipe, which is to configure remote logging in Airflow, we will use the same DAG as the *Creating basic logs in Airflow* recipe. However, feel free to create another DAG with a different name but the same code.

How to do it...

Here are the steps to perform this recipe:

1.   First, let's create a programmatic user in our AWS account. Airflow will use this user to authenticate on AWS and will be able to write the logs. On your AWS console, select **IAM services**, and you will be redirected to a page similar to this:

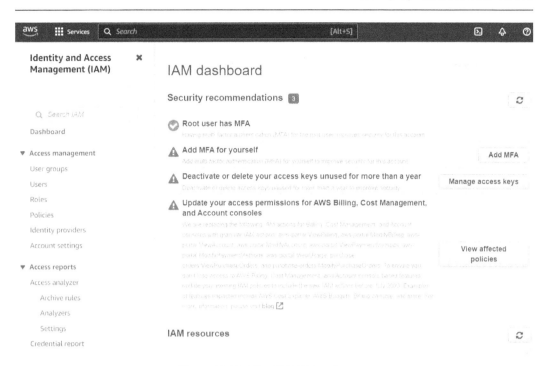

Figure 10.10 – The AWS IAM main page

2. Since this is a test account with a strict purpose, I will ignore the alerts on the IAM dashboard.

3. Then, select **Users** and **Add users**, as shown here:

Figure 10.11 – The AWS IAM Users page

On the **Create user** page, insert a username that is easy to remember, as shown here:

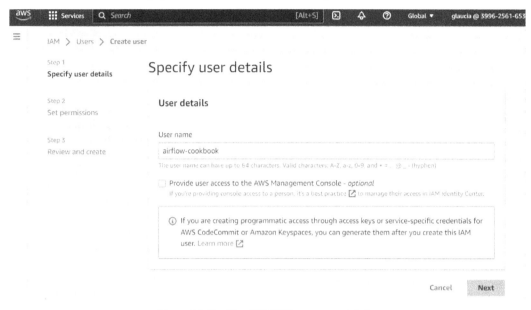

Figure 10.12 – The AWS IAM new user details

Leave the checkbox unmarked and select **Next** to add the access policies.

4.   On the **Set permissions** page, select **Attach policies directly** and then look for **AmazonS3FullAccess** in the **Permission policies** checkbox:

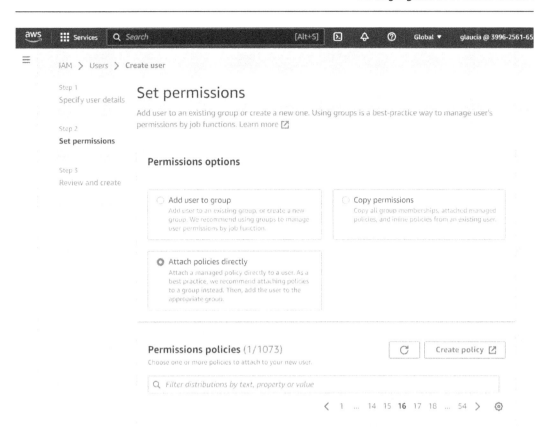

Figure 10.13 – AWS IAM set permissions for user creation

Since this is a testing exercise, we can use full access to the S3 resource. However, remember to attach specific policies to access the resources in a production environment.

Select **Next** and then click on the **Create user** button.

5. Now, retrieve the access key by selecting the user you created, go to **Security credentials**, and scroll down until you see the **Access keys** box. Then, create a new one and save the CSV file in an easily accessible place:

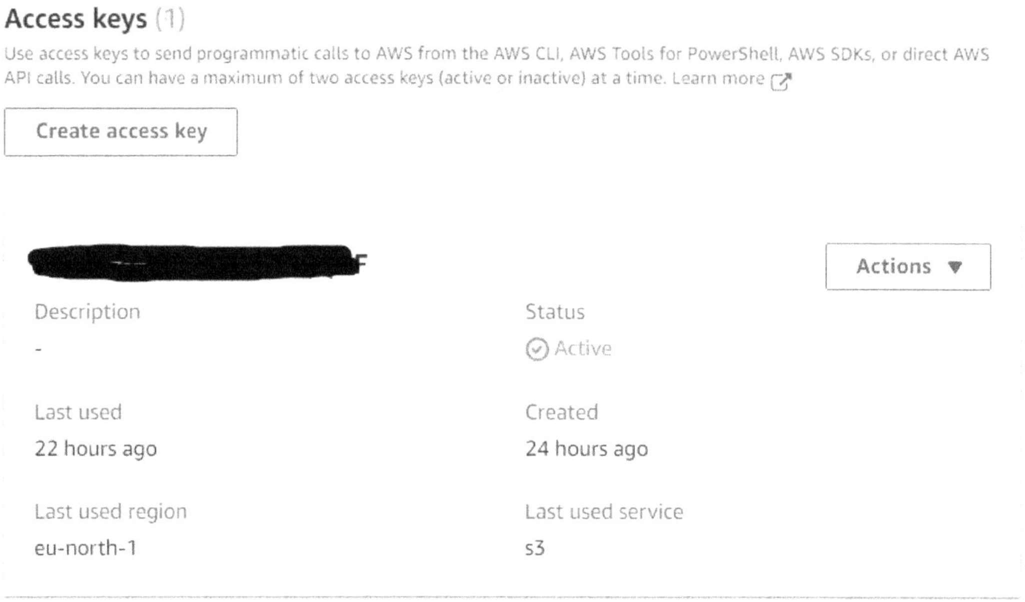

Access keys (1)

Use access keys to send programmatic calls to AWS from the AWS CLI, AWS Tools for PowerShell, AWS SDKs, or direct AWS API calls. You can have a maximum of two access keys (active or inactive) at a time. Learn more ☑

> Create access key

| | Actions ▼ |

Description	Status
-	⊘ Active

Last used	Created
22 hours ago	24 hours ago

Last used region	Last used service
eu-north-1	s3

Figure 10.14 – Access key creation for a user

6. Now, back in Airflow, let's configure the connection between Airflow and our AWS account.

    Create a new connection using the Airflow UI, and in the **Connection Type** field, select **Amazon S3**. In the **Extra** field, insert the following line with the credentials retrieved in *step 4*:

    ```
    {"aws_access_key_id": "your_key", "aws_secret_access_key":
    "your_secret"}
    ```

Your page will look like the following:

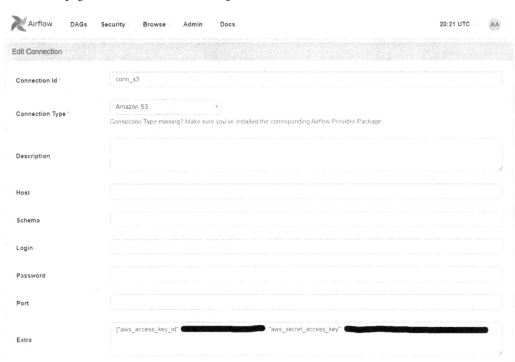

Figure 10.15 – The Airflow UI on adding a new AWS S3 connector

Save it, and open your code editor in your Airflow directory.

7. Now, let's add the configurations to our `airflow.cfg` file. If you are using Docker to host Airflow, add the following lines to your `docker-compose.yaml file`, under the environment settings:

```
AIRFLOW__LOGGING__REMOTE_LOGGING: "True"
AIRFLOW__LOGGING__REMOTE_BASE_LOG_FOLDER: "s3://airflow-
cookbook"
AIRFLOW__LOGGING__REMOTE_LOG_CONN_ID: conn_s3
AIRFLOW__LOGGING__ENCRYPT_S3_LOGS: "False"
```

Your `docker-compose.yaml` file will look similar to this:

```
version: '3'
x-airflow-common:
  &airflow-common
  # In order to add custom dependencies or upgrade provider packages you can use your extended image.
  # Comment the image line, place your Dockerfile in the directory where you placed the docker-compos
  # and uncomment the "build" line below, Then run `docker-compose build` to build the images.
  image: ${AIRFLOW_IMAGE_NAME:-apache/airflow:2.3.0}
  # build: .
  environment:
    &airflow-common-env
    AIRFLOW__CORE__EXECUTOR: CeleryExecutor
    AIRFLOW__DATABASE__SQL_ALCHEMY_CONN: postgresql+psycopg2://airflow:airflow@postgres/airflow
    # For backward compatibility, with Airflow <2.3
    AIRFLOW__CORE__SQL_ALCHEMY_CONN: postgresql+psycopg2://airflow:airflow@postgres/airflow
    AIRFLOW__CELERY__RESULT_BACKEND: db+postgresql://airflow:airflow@postgres/airflow
    AIRFLOW__CELERY__BROKER_URL: redis://:@redis:6379/0
    AIRFLOW__CORE__FERNET_KEY: ''
    AIRFLOW__CORE__DAGS_ARE_PAUSED_AT_CREATION: 'true'
    AIRFLOW__CORE__LOAD_EXAMPLES: 'true'
    AIRFLOW__API__AUTH_BACKENDS: 'airflow.api.auth.backend.basic_auth'
    # Remote logging configuration
    AIRFLOW__LOGGING__REMOTE_LOGGING: "True"
    AIRFLOW__LOGGING__REMOTE_BASE_LOG_FOLDER: "s3://airflow-cookbook"
    AIRFLOW__LOGGING__REMOTE_LOG_CONN_ID: conn_s3
    AIRFLOW__LOGGING__ENCRYPT_S3_LOGS: "False"
    PIP_ADDITIONAL_REQUIREMENTS: ${_PIP_ADDITIONAL_REQUIREMENTS:-apache-airflow-providers-mongo}
```

Figure 10.16 – Remote logging configuration in docker-compose.yaml

If you installed Airflow directly on your local machine, you can instantly change the `airflow.cfg` file. Change the following lines in `airflow.cfg` and save it:

```
[logging]
# Users must supply a remote location URL (starting with either
's3://...') and an Airflow connection
# id that provides access to the storage location.
remote_logging = True
remote_base_log_folder = s3://airflow-cookbook
remote_log_conn_id = conn_s3
# Use server-side encryption for logs stored in S3
encrypt_s3_logs = False
```

8. After the preceding changes, restart your Airflow application.

9. With your refreshed Airflow, run `basic_logging_dag` and open your AWS S3. Select the bucket you created in the *Getting ready* section, and you should see a new object inside of it, as follows:

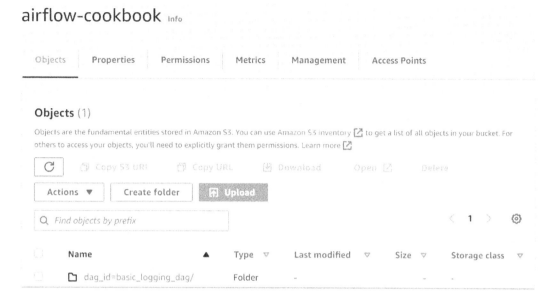

Figure 10.17 – The AWS S3 airflow-cookbook bucket objects

10. Then, select the object created, and you should be able to see more folders related to the tasks executed, as follows:

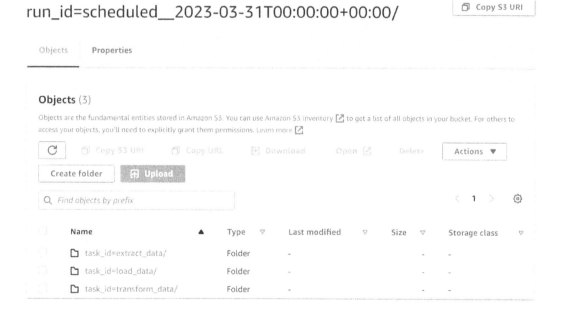

Figure 10.18 – AWS S3 airflow-cookbook showing the remote logs

11. Finally, if you select one of the folders, you will see the same file you saw in the *Creating basic logs in Airflow* recipe. We successfully wrote logs in a remote location!

How it works...

If you look at this recipe overall, it may seem considerable work. However, remember that we are making a configuration from zero, which generally takes time. Since we are somewhat used to creating an AWS S3 bucket and executing DAGs (see *Chapter 2* and *Chapter 9*, respectively), let's focus on setting the remote log configurations.

Our first action started with creating a connection in Airflow using the access keys generated on AWS. This step is required because, internally, Airflow will use those keys to authenticate in AWS and prove its identity.

Then, we changed the following Airflow configurations as follows:

```
AIRFLOW__LOGGING__REMOTE_LOGGING: "True"
AIRFLOW__LOGGING__REMOTE_BASE_LOG_FOLDER: "s3://airflow-cookbook"
AIRFLOW__LOGGING__REMOTE_LOG_CONN_ID: conn_s3
AIRFLOW__LOGGING__ENCRYPT_S3_LOGS: "False"
```

The two first lines are string configurations to set on Airflow whether remote logging is enabled and which bucket path will be used. The last two lines are related to the name of the connection we created on the **Connection** page in Airflow UI and whether we will encrypt the log messages. This last item must be set as `True` if we handle sensitive information.

After restarting Airflow, the configurations will be reflected in our application, and by executing a DAG, we can already see the logs written in the S3 bucket.

As mentioned in the introduction of this recipe, this type of configuration is beneficial not only in big projects but also as a good practice when using Airflow, allowing developers to debug or retrieve information about code output without accessing the cluster or server.

Here, we covered an example using AWS S3, but it is also possible to use **Google Cloud Storage** or **Azure Blog Storage**. You can read more here: `https://airflow.apache.org/docs/apache-airflow/1.10.13/howto/write-logs.html`.

Note

If you don't want to use remote logging anymore, you can simply remove the environment variables from your `docker-compose.yaml` or set REMOTE_LOGGING back to `False`.

See also

You can read more about remote logging in S3 on the Apache Airflow official documentation page here: `https://airflow.apache.org/docs/apache-airflow-providers-amazon/stable/logging/s3-task-handler.html`.

Configuring logs in airflow.cfg

We had our first contact with the `airflow.cfg` file in the *Storing log files in a remote location* recipe. At a glance, we saw how powerful and handy this configuration file is. There are many ways to customize and improve Airflow just by editing it.

This exercise will teach how you to enhance your logs by setting applicable configurations in the `airflow.cfg` file.

Getting ready

Refer to the *Technical requirements* section for this recipe, since we will handle it with the same technology.

Airflow DAG code

To avoid redundancy and focus on the goal of this recipe, which is to configure remote logging in Airflow, we will use the same DAG as the *Creating basic logs in Airflow* recipe. However, feel free to create another DAG with a different name but the same code.

How to do it...

Since we will use the same DAG code from *Creating basic logs in Airflow*, let's jump right to the required configuration to format our logs:

1. Let's begin by setting the configuration in our `docker-compose.yaml`. In the environment section, insert the following line and save the file:

    ```
    AIRFLOW__LOGGING__LOG_FORMAT: "[%(asctime)s] [ %(process)s -
    %(name)s ] {%(filename)s:%(lineno)d} %(levelname)s - %(message)
    s"
    ```

Your `docker-compose` file should look like this:

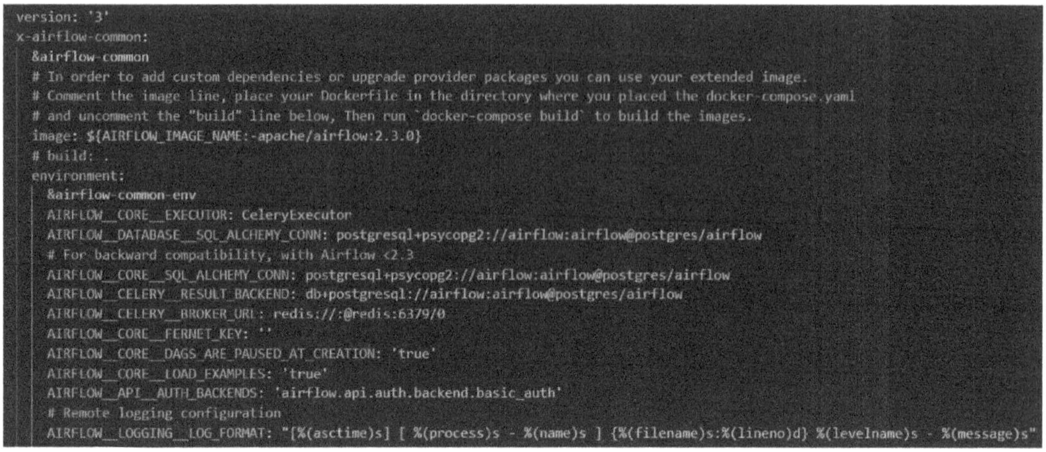

```
version: '3'
x-airflow-common:
  &airflow-common
  # In order to add custom dependencies or upgrade provider packages you can use your extended image.
  # Comment the image line, place your Dockerfile in the directory where you placed the docker-compose.yaml
  # and uncomment the "build" line below, Then run `docker-compose build` to build the images.
  image: ${AIRFLOW_IMAGE_NAME:-apache/airflow:2.3.0}
  # build: .
  environment:
    &airflow-common-env
    AIRFLOW__CORE__EXECUTOR: CeleryExecutor
    AIRFLOW__DATABASE__SQL_ALCHEMY_CONN: postgresql+psycopg2://airflow:airflow@postgres/airflow
    # For backward compatibility, with Airflow <2.3
    AIRFLOW__CORE__SQL_ALCHEMY_CONN: postgresql+psycopg2://airflow:airflow@postgres/airflow
    AIRFLOW__CELERY__RESULT_BACKEND: db+postgresql://airflow:airflow@postgres/airflow
    AIRFLOW__CELERY__BROKER_URL: redis://:@redis:6379/0
    AIRFLOW__CORE__FERNET_KEY: ''
    AIRFLOW__CORE__DAGS_ARE_PAUSED_AT_CREATION: 'true'
    AIRFLOW__CORE__LOAD_EXAMPLES: 'true'
    AIRFLOW__API__AUTH_BACKENDS: 'airflow.api.auth.backend.basic_auth'
    # Remote logging configuration
    AIRFLOW__LOGGING__LOG_FORMAT: "[%(asctime)s] [ %(process)s - %(name)s ] {%(filename)s:%(lineno)d} %(levelname)s - %(message)s"
```

Figure 10.19 – Formatting log configuration in docker-compose.yaml

If you directly edit the `airflow.cfg` file, search for the `log_format` variable, and change it to the following line:

```
log_format = [%%(asctime)s] [ %%(process)s - %%(name)s ]
{%%(filename)s:%%(lineno)d} %%(levelname)s - %%(message)s
```

Your code will look like this:

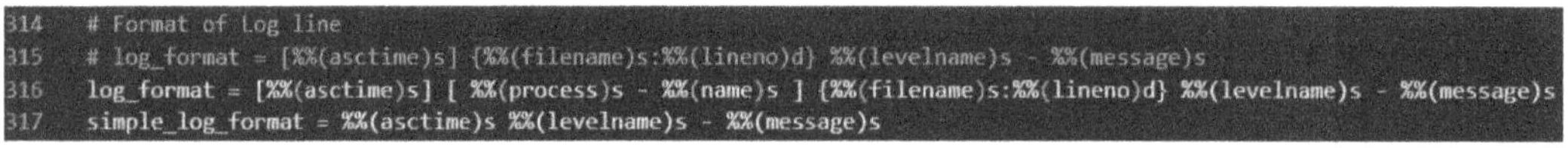

```
314    # Format of Log line
315    # log_format = [%%(asctime)s] {%%(filename)s:%%(lineno)d} %%(levelname)s - %%(message)s
316    log_format = [%%(asctime)s] [ %%(process)s - %%(name)s ] {%%(filename)s:%%(lineno)d} %%(levelname)s - %%(message)s
317    simple_log_format = %%(asctime)s %%(levelname)s - %%(message)s
```

Figure 10.20 – log_format inside airflow.cfg

Save it, and go to the next step.

We added a few more items in the log line, which we will cover later.

Note

Be very attentive here. In the `airflow.cfg` file, the % character is doubled, unlike in the `docker-compose` file.

---

2. Now, let's restart Airflow. You can do it by stopping the Docker container and rerunning it with the following commands:

```
$ docker-compose stop      # Or press Crtl-C
$ docker-compose up
```

3. Then, let's head up to the Airflow UI and run our DAG called `basic_logging_dag`. On the DAG page, look in the top-right corner and select the play button (depicted by an arrow), followed by **Trigger DAG**, as follows:

Figure 10.21 – basic_logging_dag trigger button on the right side of the page

The DAG will start to run immediately.

4. Now, let's see the logs generated by one task. I will pick the `extract_data` task, and the log will look like this:

```
[2023-04-01 21:14:52,932] [ 98 - airflow.task ] {basic_logging_dag.py:26} INFO - Let's extract data
[2023-04-01 21:14:52,932] [ 98 - airflow.task.operators ] {python.py:173} INFO - Done. Returned value was: None
```

Figure 10.22 – The formatted log output for extract_data task

If you look closely, you will see that we now have the process number displayed on the output.

> **Note**
> If you opt to maintain continuity from the last recipe, *Storing log files in a remote location*, remember that your logs are stored in a remote location.

How it works...

As we can see, altering any logging information is simple, since Airflow uses the Python logging library behind the scenes. Now, let's take a look at our output:

```
[2023-04-01 21:14:52,932] [ 98 - airflow.task ] {basic_logging_dag.py:26} INFO - Let's extract data
[2023-04-01 21:14:52,932] [ 98 - airflow.task.operators ] {python.py:173} INFO - Done. Returned value was: None
```

Figure 10.23 – The formatted log output for the extract_data task

As you can see, before the process name (for example, `airflow.task`), we also have the number of the running process. It can be helpful information when running multiple processes simultaneously, allowing us to understand which one is taking longer to complete and what is running.

Let's look at the code we inserted:

```
AIRFLOW__LOGGING__LOG_FORMAT: "[%(asctime)s] [ %(process)s - %(name)s
] {%(filename)s:%(lineno)d} %(levelname)s - %(message)s"
```

As you can see, variables such as `asctime`, `process`, and `filename` are identical to the ones we saw in *Chapter 8*. Also, since a core Python function operates behind the scenes, we can add more information based on the allowed attributes. You can find the list here: `https://docs.python.org/3/library/logging.html#logrecord-attributes`.

Going deeper in airflow.cfg

Now, let's go deeper into Airflow configurations. As you can observe, Airflow resources are orchestrated by the `airflow.cfg` file. Using a single file, we can determine how to send email notifications (we will cover this in the *Using notifications operators* recipe), when DAGs will reflect a code change, how logs will be displayed, and so on.

It is also possible to set these configurations by exporting environment variables, and this has priority over the configuration setting on `airflow.cfg`. This prioritization happens because, internally, Airflow translates the content from `airflow.cfg` to environment variables, broadly speaking. You can read more here: `https://airflow.apache.org/docs/apache-airflow/stable/cli-and-env-variables-ref.html#environment-variable`.

Let's look at the logging configuration in the Airflow **REFERENCES** section. We can see many other customization possibilities, such as coloring, a specific format for DAG processors, and extra logs for third-party applications, as shown here:

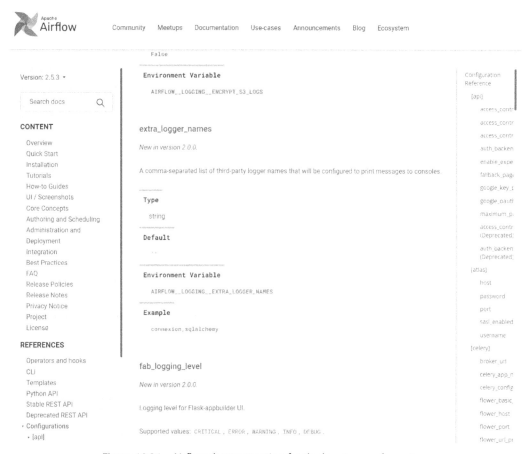

Figure 10.24 – Airflow documentation for the logging configuration

The fantastic part of this documentation is that we have references to configure directly in `airflow.cfg` or environment variables. You can see the complete reference list here: `https://airflow.apache.org/docs/apache-airflow/stable/configurations-ref.html#logging`.

After we get used to the Airflow dynamics, testing new configurations or formats is straightforward, especially when we have a testing server to do so. However, simultaneously, we need to be cautious when changing anything internally; otherwise, we can impair our whole application.

There's more...

In *step 1*, we mentioned avoiding the use of double `%` characters when setting the variables in `docker-compose` – let's now cover this!

The `string` variable we pass for `docker-compose` will be read by an internal Python logging function, which will not recognize the double % pattern. Instead, it will understand the default format for the logs in Airflow needs to be equal to that string variable, and all the DAG logs will look like this:

Log by attempts

1   2   3   [4]   5   6   7   8   Jump To End   Toggle Wrap   Download

```
*** Reading remote log from s3://airflow-cookbook/dag_id=basic_logging_dag/run_id=scheduled__2023-03-31T00:00+00:00/task_id=extract_data/attempt=4.log.
[%(asctime)s] [ %(process)s - %(name)s ] {%(filename)s:%(lineno)d} %(levelname)s - %(message)s
[%(asctime)s] [ %(process)s - %(name)s ] {%(filename)s:%(lineno)d} %(levelname)s - %(message)s
[%(asctime)s] [ %(process)s - %(name)s ] {%(filename)s:%(lineno)d} %(levelname)s - %(message)s
[%(asctime)s] [ %(process)s - %(name)s ] {%(filename)s:%(lineno)d} %(levelname)s - %(message)s
[%(asctime)s] [ %(process)s - %(name)s ] {%(filename)s:%(lineno)d} %(levelname)s - %(message)s
[%(asctime)s] [ %(process)s - %(name)s ] {%(filename)s:%(lineno)d} %(levelname)s - %(message)s
[%(asctime)s] [ %(process)s - %(name)s ] {%(filename)s:%(lineno)d} %(levelname)s - %(message)s
[%(asctime)s] [ %(process)s - %(name)s ] {%(filename)s:%(lineno)d} %(levelname)s - %(message)s
[%(asctime)s] [ %(process)s - %(name)s ] {%(filename)s:%(lineno)d} %(levelname)s - %(message)s
[%(asctime)s] [ %(process)s - %(name)s ] {%(filename)s:%(lineno)d} %(levelname)s - %(message)s
[%(asctime)s] [ %(process)s - %(name)s ] {%(filename)s:%(lineno)d} %(levelname)s - %(message)s
[%(asctime)s] [ %(process)s - %(name)s ] {%(filename)s:%(lineno)d} %(levelname)s - %(message)s
[%(asctime)s] [ %(process)s - %(name)s ] {%(filename)s:%(lineno)d} %(levelname)s - %(message)s
```

Figure 10.25 – An error when the environment variable for log_format is not correctly set

Now, inside the `airflow.cfg` file, the double % character is a Bash format pattern that works like a modulo operator.

See also

See the whole list of configurations for Airflow here: `https://airflow.apache.org/docs/apache-airflow/stable/configurations-ref.html`.

Designing advanced monitoring

After spending some time learning and practicing logging concepts, we can advance a little more in the subject of monitoring. We can monitor results from all our logging collection work and generate insightful monitoring dashboards and alerts, with the right monitoring message stored.

In this recipe, we will cover the Airflow metrics integrated with StatsD, a platform that collects system statistics, and their purpose to help us achieve a mature pipeline.

Getting ready

This exercise will focus on bringing clarity to the Airflow monitoring metrics and how to build a robust architecture to structure it.

As a requirement for this recipe, it is vital to keep in mind the following basic Airflow architecture:

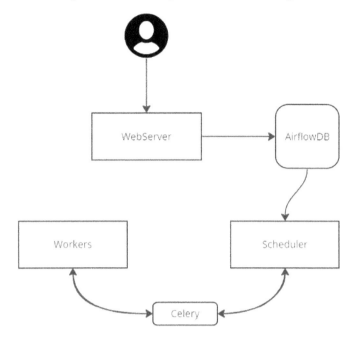

Figure 10.26 – An Airflow high-level architecture diagram

Airflow components, from a high-level perspective, are composed of the following:

- A **web server**, where we can access the Airflow UI.
- A relational database to store metadata and other helpful information for use in the DAGs or tasks. To keep it simple, we will work with just one type of database; however, there can be more than one.
- The **scheduler**, which will consult the information inside the database to send it to the workers.
- A **Celery** application, responsible for queueing the requests sent from the scheduler and the workers.
- The **workers**, which will execute the DAG and tasks.

With this in mind, we can proceed to the next section.

How to do it...

Let's see the main items to design advanced monitoring:

- **Counters**: As the name suggests, this metric will provide information about the counts of actions inside Airflow. This metric provides a count of running tasks, failed tasks, and so on. In the following figure, you can see some examples:

Figure 10.27 – A list of counter metric examples to monitor Airflow workflows

- **Timers**: This metric tells us how long a task or DAG takes to complete or load a file. In the following figure, you can see more:

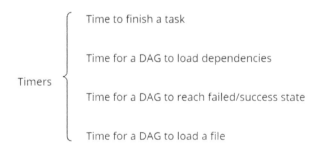

Figure 10.28 – A list of timer examples to monitor Airflow workflows

- **Gauges**: Finally, the last metric type gives us a more visual overview. Gauges use timers or counters metrics to illustrate whether we are reaching a defined threshold. In the following figure, there are some examples of gauges:

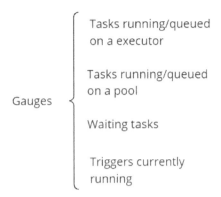

Figure 10.29 – A list of gauge examples to be used to monitor Airflow

With the metrics defined and on our radar, we can proceed with the architecture design to integrate it.

- **StatsD**: Now, let's add **StatsD** to the architecture drawing we saw in the *Getting ready* section. You will have something like this:

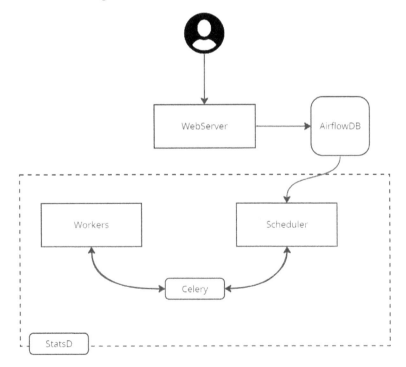

Figure 10.30 – StatsD integration and coverage for the Airflow components architecture

StatsD can collect the metrics from all the components inside the dotted rectangle and direct them to a monitoring tool.

- **Prometheus and Grafana**: Then, we can plug StatsD into Prometheus, which serves as one of Grafana's data sources. Adding these tools into our architecture will look something like this:

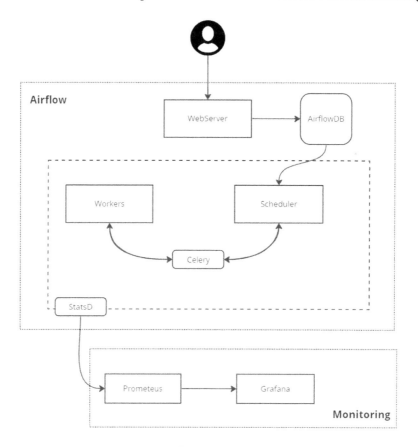

Figure 10.31 – A Prometheus and Grafana integration with StatsD and Airflow diagram

Now, let's understand the components behind this architecture.

How it works...

Let's start understanding what StatsD is. StatsD is a daemon developed by the Etsy company to aggregate and collect application metrics. Generally, any application can send metrics using a simple protocol, such as **User Datagram Protocol** (**UDP**). With this protocol, the sender doesn't need to wait for a response from StatsD, making the process simple. After listening and aggregating data for some time, StatsD will send the metrics to output storage, which is Prometheus.

The StatsD integration and installation can be done using the following command:

```
pip install 'apache-airflow[statsd]'
```

If you want to know more about it, you can refer to the Airflow documentation here: `https://airflow.apache.org/docs/apache-airflow/2.5.1/administration-and-deployment/logging-monitoring/metrics.html#counters`.

Then, Prometheus and Grafana will gather the metrics and transform them into a more visual resource. You don't need to worry about this now; we will learn more about it in *Chapter 12*.

For each metric we saw in the three first steps in the *How to do it...* section, we can set a threshold to trigger an alert when it has trespassed. All the metrics are presented in the *How to do it...* section, and some more can be found here: `https://airflow.apache.org/docs/apache-airflow/2.5.1/administration-and-deployment/logging-monitoring/metrics.html#counters`.

There's more...

Besides StatsD, there are other tools we can plug into Airflow to track specific metrics or statuses. For example, for a deep error track, we can use **Sentry**, a specialized tool used by IT operations teams to provide support and insights. You can learn more about this integration here: `https://airflow.apache.org/docs/apache-airflow/stable/administration-and-deployment/logging-monitoring/errors.html`.

On the other hand, if tracking users' activities is a concern, it is possible to integrate Airflow with Google Analytics. You can learn more here: `https://airflow.apache.org/docs/apache-airflow/stable/administration-and-deployment/logging-monitoring/tracking-user-activity.html`.

See also

- Learn more about Airflow architecture here: `https://airflow.apache.org/docs/apache-airflow/stable/administration-and-deployment/logging-monitoring/logging-architecture.html`
- More information about StatsD is here: `https://www.datadoghq.com/blog/statsd/`

Using notification operators

So far, we have focused on ensuring that code is well logged and has enough information to provide valid monitoring. Nevertheless, the purpose of having mature and structured pipelines is to avoid the necessity of manual intervention. With busy agendas and other projects, it is hard to constantly look at monitoring dashboards to check whether everything is fine.

Thankfully, Airflow also has native operators to trigger alerts depending on their configured situation. In this recipe, we will configure an email operator to trigger a message every time a pipeline succeeds or fails, allowing us to remediate the problem rapidly.

Getting ready

Refer to the *Technical requirements* section for this recipe, since we will handle it with the same technology.

In addition to that, you need to create an app password for your Google account. This password will allow our application to authenticate and use the **Simple Mail Transfer Protocol** (**SMTP**) host from Google to trigger emails. You can generate the app password in your Google account at the following link: `https://security.google.com/settings/security/apppasswords`.

Once you access the link, you will be asked to authenticate using your Google credentials, and a new page will appear, similar to the following:

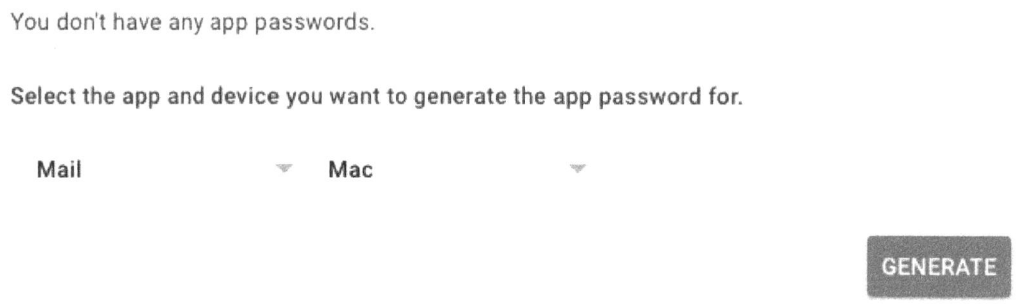

You don't have any app passwords.

Select the app and device you want to generate the app password for.

Mail	Mac

GENERATE

Figure 10.32 – The Google app password generation page

In the first box, select **Mail**, and in the second box, select the device that will use the app password. Since I am using a Macbook, I will select **Mac**, as shown in the preceding screenshot. Then, click on **GENERATE**.

A window similar to the following will appear:

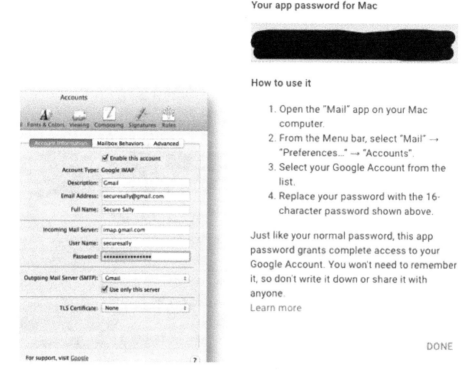

Figure 10.33 – The Google generated app password pop-up window

Follow the steps on the page and save the password in a place you can remember.

Airflow DAG code

To avoid redundancy and focus on the goal of this recipe, which is to configure remote logging in Airflow, we will use the same DAG as the *Creating basic logs in Airflow* recipe. However, feel free to create another DAG with a different name but the same code.

Nonetheless, you can always find the final code in the GitHub repository here:

```
https://github.com/PacktPublishing/Data-Ingestion-with-Python-
Cookbook/tree/main/Chapter_10/Using_notifications_operators
```

How to do it...

Perform the following steps to try this recipe:

1. Let's start by configuring the SMTP server in Airflow. Insert the following lines in your `docker-compose.yaml` file under the environment section:

```
# SMTP settings
AIRFLOW__SMTP__SMTP_HOST: "smtp.gmail.com"
AIRFLOW__SMTP__SMTP_USER: "your_email_here"
AIRFLOW__SMTP__SMTP_PASSWORD: "your_app_password_here"
AIRFLOW__SMTP__SMTP_PORT: 587
```

Your file should look like this:

```
version: '3'
x-airflow-common:
  &airflow-common
  # In order to add custom dependencies or upgrade provider packages you can use your extended image.
  # Comment the image line, place your Dockerfile in the directory where you placed the docker-compose.y
  # and uncomment the "build" line below, Then run `docker-compose build` to build the images.
  image: ${AIRFLOW_IMAGE_NAME:-apache/airflow:2.3.0}
  # build: .
  environment:
    &airflow-common-env
    AIRFLOW__CORE__EXECUTOR: CeleryExecutor
    AIRFLOW__DATABASE__SQL_ALCHEMY_CONN: postgresql+psycopg2://airflow:airflow@postgres/airflow
    # For backward compatibility, with Airflow <2.3
    AIRFLOW__CORE__SQL_ALCHEMY_CONN: postgresql+psycopg2://airflow:airflow@postgres/airflow
    AIRFLOW__CELERY__RESULT_BACKEND: db+postgresql://airflow:airflow@postgres/airflow
    AIRFLOW__CELERY__BROKER_URL: redis://:@redis:6379/0
    AIRFLOW__CORE__FERNET_KEY: ''
    AIRFLOW__CORE__DAGS_ARE_PAUSED_AT_CREATION: 'true'
    AIRFLOW__CORE__LOAD_EXAMPLES: 'true'
    AIRFLOW__API__AUTH_BACKENDS: 'airflow.api.auth.backend.basic_auth'
    _PIP_ADDITIONAL_REQUIREMENTS: ${_PIP_ADDITIONAL_REQUIREMENTS:-apache-airflow-providers-mongo apache-
    # SMTP settings
    AIRFLOW__SMTP__SMTP_HOST: "smtp.gmail.com"
    AIRFLOW__SMTP__SMTP_USER: "               "
    AIRFLOW__SMTP__SMTP_PASSWORD: "          "
    AIRFLOW__SMTP__SMTP_PORT: 587
```

Figure 10.34 – docker-compose.yaml with SMTP environment variables

If you directly edit the `airflow.cfg` file, edit the following lines:

```
[smtp]
# If you want airflow to send emails on retries, failure, and
you want to use
# the airflow.utils.email.send_email_smtp function, you have to
configure an
# smtp server here
```

```
smtp_host = smtp.gmail.com
smtp_starttls = True
smtp_ssl = False
# Example: smtp_user = airflow
smtp_user = your_email_here
# Example: smtp_password = airflow
smtp_password = your_app_password_here
smtp_port = 587
smtp_mail_from = airflow@example.com
smtp_timeout = 30
smtp_retry_limit = 5
```

Don't forget to restart Airflow after these configurations are saved.

2. Now, let's edit our `basic_logging_dag` DAG to allow it to send emails using `EmailOperator`. Let's add to our imports the following line:

```
from airflow.operators.email import EmailOperator
```

The imports will be organized like this:

```
from airflow import DAG
from airflow.operators.python_operator import PythonOperator
from airflow.operators.email import EmailOperator
from datetime import datetime, timedelta
import logging

# basic_logging_dag DAG code
# ...
```

3. In `default_args`, we will add three new parameters – `email`, `email_on_failure`, and `email_on_retry`. You can see here what it looks like:

```
# basic_logging_dag DAG imports above this line

default_args = {
    'owner': 'airflow',
    'depends_on_past': False,
    'start_date': datetime(2023, 4, 1),
    'email': ['sample@gmail.com'],
    'email_on_failure': True,
    'email_on_retry': True,
    'retries': 1,
    'retry_delay': timedelta(minutes=5)
```

```
    }

    # basic_logging_dag DAG code
    # …
```

You don't need to worry about these new parameters for now. We will cover them in the *How it works…* section.

4. Then, let's add a new task to our DAG called `success_task`. If all the other tasks are successful, this one will trigger `EmailOperator` to alert us. Add the following code to the `basic_logging_dag` script:

```
success_task = EmailOperator(
    task_id="success_task",
    to= "g.esppen@gmail.com",
    subject="The pipeline finished successfully!",
    html_content="<h2> Hello World! </h2>",
    dag=dag
)
```

5. Finally, at the end of your script, let's add the workflow:

```
extract_task >> transform_task >> load_task >> success_task
```

Don't forget that you can always check how the final code looks here: `https://github.com/PacktPublishing/Data-Ingestion-with-Python-Cookbook/tree/main/Chapter_10/Using_noti%EF%AC%81cations_operators`

6. If you check your DAG graph, you can see that a new task called `success_task` appears. It shows our operator is ready to be used. Let's trigger our DAG by selecting the play button in the top-right corner, as we did in *step 3* of the *Configuring logs in airflow.cfg* recipe.

Your Airflow UI should look like this:

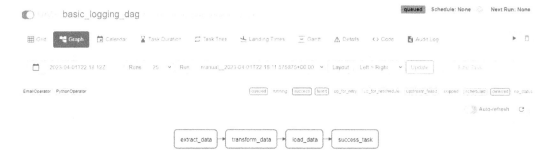

Figure 10.35 – basic_logging_dag showing successful runs for all the tasks

7. Then, let's check our email. If everything is well configured, you should see an email similar to the following:

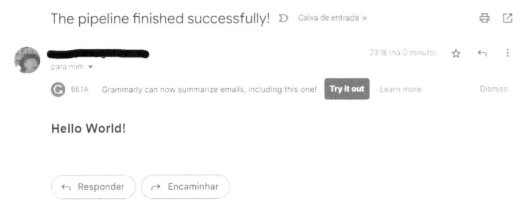

Figure 10.36 – An email with a Hello World! Message, indicating that success_task worked

Our `EmailOperator` works exactly as expected!

How it works...

Let's start explaining the code by defining what an SMTP server is. An SMTP server is a key component of an email system that enables the transmission of email messages between servers and from clients to servers.

In our case, Google works both as a sender and receiver. We borrow a Gmail host to help send an email from our local machine. However, you don't need to worry about this when working on a company project; your IT operations team will take care of it.

Now, back to Airflow – once we understand how the SMTP works, its configuration is straightforward. Consulting the reference page for the configurations in Airflow (`https://airflow.apache.org/docs/apache-airflow/stable/configurations-ref.html`), we can see that there is a section dedicated to SMTP, as you can see here:

[smtp]

If you want airflow to send emails on retries, failure, and you want to use the airflow.utils.email.send_email_smtp function, you have to configure an smtp server here

smtp_host

Type

string

Default

localhost

Environment Variable

AIRFLOW__SMTP__SMTP_HOST

Figure 10.37 – The Airflow documentation page for the SMTP environment variables

Then, all we needed to do was to set the required parameters to allow the connection between the host (smtp.gmail.com) and Airflow, as you can see here:

```
# SMTP settings
AIRFLOW__SMTP__SMTP_HOST: "smtp.gmail.com"
AIRFLOW__SMTP__SMTP_USER:            "
AIRFLOW__SMTP__SMTP_PASSWORD:            "
AIRFLOW__SMTP__SMTP_PORT: 587
```

Figure 10.38 – A close look at the docker-compose.yaml SMTP settings

Once this step is completed, we will go to our DAG and declare EmailOperator, as shown in the following code:

```
success_task = EmailOperator(
    task_id="success_task",
    to="g.esppen@gmail.com",
    subject="The pipeline finished successfully!",
    html_content="<h2> Hello World! </h2>",
    dag=dag
)
```

The parameters of the email are very intuitive and can be set accordingly to whatever is needed. If we delve deeper, we can see that there are plenty of possibilities to make those fields' values more abstract to adapt to different function results.

It is also possible to use a formatted email template in `html_content` and even attach a complete error or log message. You can see more of the allowed parameters here: `https://airflow.apache.org/docs/apache-airflow/stable/_api/airflow/operators/email/index.html`.

In our case, this operator was triggered when all tasks successfully ran. But what about if there is an error? Let's go back to *step 3* and see `default_args`:

```
default_args = {
    'owner': 'airflow',
    'depends_on_past': False,
    'start_date': datetime(2023, 4, 1),
    'email': ['sample@gmail.com'],
    'email_on_failure': True,
    'email_on_retry': True,
    'retries': 1,
    'retry_delay': timedelta(minutes=5)
}
```

The two new parameters added (`email_on_failure` and `email_on_retry`) address scenarios where the DAG failed or retries a task. The values inside the `email` parameter list are the recipients of these emails.

A default email triggered by an error message looks like this:

Figure 10.39 – The Airflow default email for error in a task instance

There's more...

The Airflow notification system is not limited to sending emails and counts, offering useful integrations with Slack, Teams, and Telegram.

TowardsDataScience has a fantastic blog post about how to integrate Airflow with Slack, and you can find it here: `https://towardsdatascience.com/automated-alerts-for-airflow-with-slack-5c6ec766a823`.

Not limited to corporate tools, Airflow also has a Discord hook: `https://airflow.apache.org/docs/apache-airflow-providers-discord/stable/_api/airflow/providers/discord/hooks/discord_webhook/index.html`.

The best advice I can give is always to look at Airflow community documentation. As an open source and active platform, there is always a new implementation to help automate and make our daily work easier.

Using SQL operators for data quality

Good **data quality** is crucial for an organization to ensure the effectiveness of its data systems. By performing quality checks within the DAG, it is possible to stop pipelines and notify stakeholders before erroneous data is introduced into a production lake or warehouse.

Although plenty of available tools in the market provide **data quality checks**, one of the most popular ways to do this is by running SQL queries. As you may have already guessed, Airflow has providers to support those operations.

This recipe will cover the data quality principal topics in the data ingestion process, pointing out the best `SQLOperator` type to run in those situations.

Getting ready

Before starting our exercise, let's create a simple **Entity Relationship Diagram** (**ERD**) for a `customers` table. You can see here how it looks:

Figure 10.40 – An example of customers table columns

And the same table is represented with its schema:

```
CREATE TABLE customers (
    customer_id INT PRIMARY KEY,
    first_name VARCHAR(50),
    last_name VARCHAR(50),
    email VARCHAR(100),
    phone_number VARCHAR(20),
    address VARCHAR(200),
    city VARCHAR(50),
    state VARCHAR(50),
    country VARCHAR(50),
    zip_code VARCHAR(20)
);
```

You don't need to worry about creating this table in a SQL database. This exercise will focus on the data quality factors to be checked, using this table as an example.

How to do it...

Here are the steps to perform this recipe:

1. Let's start by defining the essential data quality checks that apply as follows:

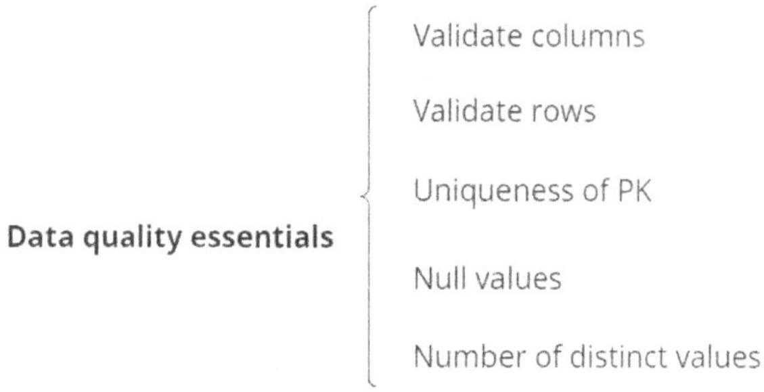

Data quality essentials

- Validate columns
- Validate rows
- Uniqueness of PK
- Null values
- Number of distinct values

Figure 10.41 – Data quality essential points

2. Let's imagine implementing it using SQLColumnCheckOperator, integrated and installed in our Airflow platform. Let's now create a simple task to check whether our table has unique IDs and whether all customers have first_name. Our example code looks like this:

```
id_username_check = SQLColumnCheckOperator(
        task_id="id_username_check",
        conn_id= my_conn,
        table=my_table,
        column_mapping={
            "customer_id": {
                "null_check": {
                    "equal_to": 0,
                    "tolerance": 0,
                },
                "distinct_check": {
                    "equal_to": 1,
                },
            },
            "first_name": {
                "null_check": {"equal_to": 0},
            },
        }
    )
```

3. Now, let's validate whether we ingest the required count of rows using SQLTableCheckOperator, as follows:

```
customer_table_rows_count = SQLTableCheckOperator(
    task_id="customer_table_rows_count",
    conn_id= my_conn,
    table=my_table,
    checks={"row_count_check": {
                "check_statement": "COUNT(*) >= 1000"
            }
        }
)
```

4. Finally, let's ensure the customers in our database have at least one order. Our example code looks like this:

```
count_orders_check = SQLColumnCheckOperator(
    task_id="check_columns",
    conn_id=my-conn,
    table=my_table,
    column_mapping={
        "MY_NUM_COL": {
            "min": {"geq_to ": 1}
        }
    }
)
```

The geq_to key stands for **great or equal to**.

How it works...

Data quality is a complex topic encompassing many variables, such as the project or company context, business models, and **Service Level Agreements** (**SLAs**) between teams. Based on this, the goal of this recipe was to offer the core concept of data quality and demonstrate how to first approach using Airflow SQLOperators.

Let's start with the essential topics in *step 1*, as follows:

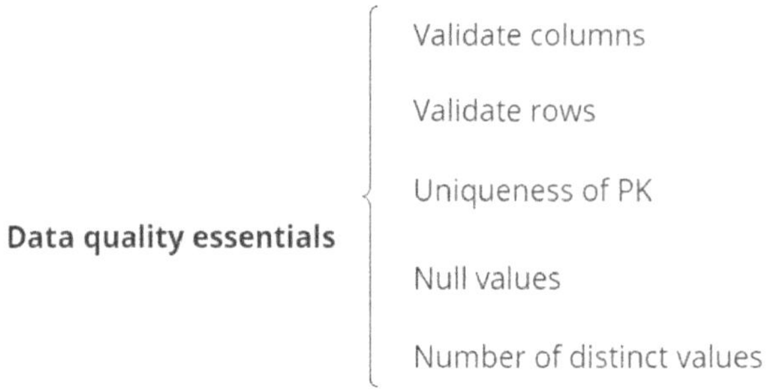

Validate columns

Validate rows

Uniqueness of PK

Data quality essentials

Null values

Number of distinct values

Figure 10.42 – Data quality essential points

In a generic scenario, those items are the principal topics to be approached and implemented. They will guarantee the minimum data reliability, based on whether the columns are the ones we expected, creating an average value for the row count, ensuring the IDs are unique, and having control of the null and distinct values in specific columns.

Using Airflow, we used the SQL approach to check data. As mentioned at the beginning of this recipe, SQL checks are widespread and popular due to their simplicity and flexibility. Unfortunately, to simulate a scenario like this, we would be required to set up a hard-working local infrastructure, and the best we could come up with is simulating the tasks in Airflow.

Here, we used two SQLOperator subtypes – SQLColumnCheckOperator and SQLTableCheckOperator. As the name suggests, the first operator is more focused on verifying the column's content by checking whether there are null or distinct values. In the case of customer_id, we verified both scenarios and only null values for first_name, as you can see here:

```
column_mapping={
        "customer_id": {
            "null_check": {
                "equal_to": 0,
                "tolerance": 0,
            },
            "distinct_check": {
                "equal_to": 1,
            },
        },
        "first_name": {
            "null_check": {"equal_to": 0},
        },
    }
```

`SQLTableCheckOperator` will perform validations across the whole table. It allows the insertion of a SQL query to make counts or other operations, as we did to validate the expected number of rows in *step 3*, as you can see in the piece of code here:

```
checks={"row_count_check": {
        "check_statement": "COUNT(*) >= 1000"
    }
}
```

However, `SQLOperator` is not limited to these two. In the Airflow documentation, you can see other examples and the complete list of accepted parameters for these functions: `https://airflow. apache.org/docs/apache-airflow/2.1.4/_api/airflow/operators/sql/ index.html#module-airflow.operators.sql`.

A fantastic operator to check out is `SQLIntervalCheckOperator`, used to validate historical data and ensure the stored information is concise.

In your data career, you will see that data quality is a daily topic and concern among teams. The best advice here is to continually search for tools and methods to improve this methodology.

There's more...

We can use additional tools to enhance our data quality checks. One of the recommended tools for this use is **GreatExpectations**, an open source platform made in Python with plenty of integrations, with resources such as Airflow, **AWS S3**, and **Databricks**.

Although it is a platform you can install in any cluster, **GreatExpectations** is expanding toward a managed cloud version. You can check more about it on the official page here: `https:// greatexpectations.io/integrations`.

See also

- *Yu Ishikawa* has a nice blog post about other checks you can do using SQL in Airflow: `https:// yu-ishikawa.medium.com/apache-airflow-as-a-data-quality-checker- 416ca7f5a3ad`

- More information about data quality in Airflow is available here: `https://docs. astronomer.io/learn/data-quality`

Further reading

- `https://www.oak-tree.tech/blog/airflow-remote-logging-s3`
- `https://airflow.apache.org/docs/apache-airflow-providers-amazon/stable/connections/aws.html#examples`
- `https://airflow.apache.org/docs/apache-airflow/stable/howto/email-config.html`
- `https://docs.astronomer.io/learn/logging`
- `https://airflow.apache.org/docs/apache-airflow/stable/administration-and-deployment/logging-monitoring/metrics.html#setup`
- `https://hevodata.com/learn/airflow-monitoring/#aam`
- `https://servian.dev/developing-5-step-data-quality-framework-with-apache-airflow-972488ddb65f`

11

Automating Your Data Ingestion Pipelines

Data sources are frequently updated, and this requires us to update our data lake. However, with multiple sources or projects, it becomes impossible to trigger data pipelines manually. Data pipeline automation makes ingesting and processing data mechanical, obviating the human actions to trigger it. The importance of automation configuration lies in the ability to streamline data flow and improve data quality, reducing errors and inconsistency.

In this chapter, we will cover how to automate the data ingestion pipelines in Airflow, along with two essential topics in data engineering, data replication and historical data ingestion, as well as best practices.

In this chapter, we will cover the following recipes:

- Scheduling daily ingestions
- Scheduling historical data ingestion
- Scheduling data replication
- Setting up the `schedule_interval` parameter
- Solving scheduling errors

Technical requirements

You can find the code from this chapter in the GitHub repository at `https://github.com/PacktPublishing/Data-Ingestion-with-Python-Cookbook/tree/main/Chapter_11`.

Installing and running Airflow

This chapter requires that Airflow is installed on your local machine. You can install it directly on your **Operating System (OS)** or use a Docker image. For more information, refer to the *Configuring Docker for Airflow* recipe in *Chapter 1*.

After following the steps described in *Chapter 1*, ensure your Airflow instance runs correctly. You can do that by checking the Airflow UI at `http://localhost:8080`.

If you are using a Docker container (as I am) to host your Airflow application, you can check its status in the terminal with the following command:

```
$ docker ps
```

Here is the status of the container:

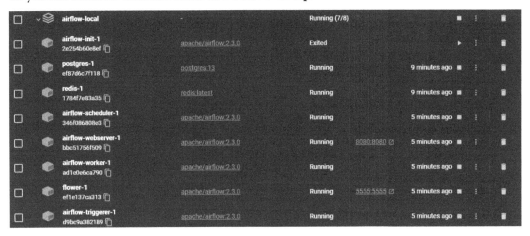

Figure 11.1 – Airflow containers running

Or you can check the container status on **Docker Desktop**:

Figure 11.2 – Docker Desktop showing Airflow running containers

Scheduling daily ingestions

Data constantly changes in our dynamic world, with new information being added every day and even every second. Therefore, it is crucial to regularly update our data lake to reflect the latest scenarios and information.

Managing multiple projects or pipelines concurrently and manually triggering them while integrating new data from various sources can be daunting. To alleviate this issue, we can rely on schedulers, and Airflow provides a straightforward solution for this purpose.

In this recipe, we will create a simple **Directed Acyclic Graph** (**DAG**) in Airflow and explore how to use its parameters to schedule a pipeline to run daily.

Getting ready

Please refer to the *Technical requirements* section for this recipe since we will handle it with the same technology mentioned here.

In this exercise, we will create a simple DAG. The structure of your Airflow folder should look like the following:

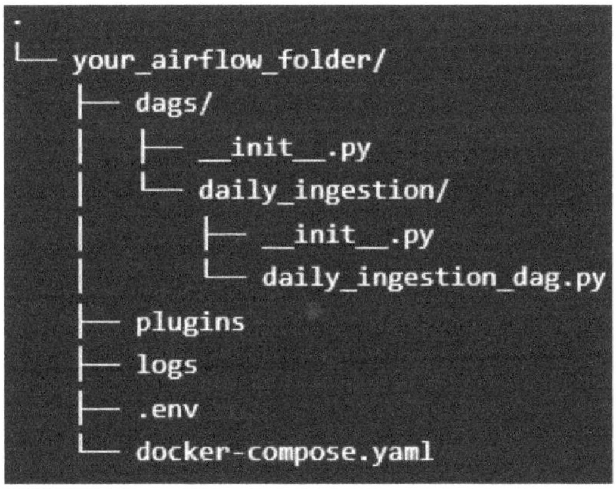

```
.
└── your_airflow_folder/
    ├── dags/
    │   ├── __init__.py
    │   └── daily_ingestion/
    │       ├── __init__.py
    │       └── daily_ingestion_dag.py
    ├── plugins
    ├── logs
    ├── .env
    └── docker-compose.yaml
```

Figure 11.3 – daily_ingestion_dag DAG folder structure

All code in this recipe will be placed inside `daily_ingestion_dag.py`. Ensure you have created the file by following the folder structure in *Figure 11.3*.

How to do it...

These are the steps for this recipe:

1. Let's start by importing the required libraries:

```
from airflow import DAG
from airflow.operators.bash import BashOperator
from datetime import datetime, timedelta
```

2. Now, we will define `default_args` for our DAG. For the `start_date` parameter, insert today's date or a few days before you are doing this exercise. For `end_date`, insert a date a few days ahead of today's date. In the end, it should look like the following:

```
default_args = {
    'owner': 'airflow',
    'depends_on_past': False,
    'email': ['airflow@example.com'],
    'email_on_failure': True,
    'email_on_retry': True,
    'retries': 1,
    'retry_delay': timedelta(minutes=5),
    'start_date': datetime(2023, 4, 12),
    'end_date': datetime(2023, 4, 30),
    'schedule_interval': '@daily
}
```

3. Then, we will define our DAG and the tasks inside it. Since we want to focus on how to schedule daily ingestion, our tasks will each be a `BashOperator` since they can execute Bash commands with simplicity, as you can see here:

```
with DAG(
    'daily_ingestion_dag',
    default_args=default_args,
    description='A simple ETL job using Bash commands',
) as dag:

    t1 = BashOperator(
            task_id="t1",
            bash_command="echo 'This is task no1 '",
        )

    t2 = BashOperator(
            task_id="t2",
            bash_command="echo 'This is task no2 '",
        )

t1 >> t2
```

4. With the DAG written, let's enable it on the Airflow UI, and the DAG should run immediately. After running, the DAG will have a SUCCESS status, as follows:

Figure 11.4 – daily_ingestion_dag DAG in the Airflow UI

If we check the logs, it will show the `echo` command output, similar to the following:

```
[2023-04-12, 19:54:38 UTC] [ 1686 - airflow.hooks.subprocess.
SubprocessHook ] {subprocess.py:74} INFO - Running command:
['bash', '-c', "echo 'This is task no2 '"]
[2023-04-12, 19:54:38 UTC] [ 1686 - airflow.hooks.subprocess.
SubprocessHook ] {subprocess.py:85} INFO - Output:
[2023-04-12, 19:54:38 UTC] [ 1686 - airflow.hooks.subprocess.
SubprocessHook ] {subprocess.py:92} INFO - This is task no2
```

5. Now, we need to ensure the DAG will run daily. To confirm this, select the **Calendar** option on your DAG page. You will see something similar to this:

Figure 11.5 – DAG's Calendar visualization in the Airflow UI

As you can see, the execution is depicted in the shaded region to the left, indicating a successful outcome (**SUCCESS**) relative to the current date. The following days, until `end_date`, are marked with a dot inside, indicating the job will run every day for the next few days.

> **Note**
>
> *Figure 11.5* shows some days when the job was executed successfully. This is shown to users how the same calendar behaves on previous executions.

How it works...

Airflow's scheduler is mainly defined by three parameters: `start_date`, `end_date`, and `schedule_interval`. These three parameters define the beginning and end of the job and the interval between executions.

Let's take a look at `default_args`:

```
default_args = {
    'owner': 'airflow',
    ...
    'start_date': datetime(2023, 4, 12),
    'end_date': datetime(2023, 4, 30),
    'schedule_interval': '@daily
}
```

Since I am writing this exercise on April 12, 2023, I set my `start_date` parameter to the same day. This will make the job retrieve information relating to April 12, and if I put it a few days before the current date, Airflow will retrieve the earlier date. Don't worry about it now; we will cover more about this in the *Scheduling historical data ingestion recipe*.

The key here is the `schedule_interval` parameter. As the name suggests, this parameter will define the periodicity or the interval of each execution, and, as you can observe, it was simply set using the `@daily` value.

The **Calendar** option on the DAG UI page is an excellent feature of Airflow 2.2 onward. This functionality allows the developers to see the next execution days for the DAG, preventing some confusion.

There's more...

The DAG parameters are not limited to the ones we have seen in this recipe. Many others are available to make the data pipeline even more automated and intelligent. Let's take a look at the following code:

```
default_args = {
    'owner': 'airflow',
    'depends_on_past': False,
    'email': ['airflow@example.com'],
    'email_on_failure': True,
    'email_on_retry': True,
    'retries': 1,
```

```
        'retry_delay': timedelta(minutes=5),
        'start_date': datetime(2023, 4, 12),
        'end_date': datetime(2023, 4, 30),
        'schedule_interval': '@daily,
        'queue': 'bash_queue',
        'pool': 'backfill',
        'priority_weight': 10
    }
```

There are three additional parameters here: `queue`, `pool`, and `priority_weight`. As we saw in *Chapter 9* and *Chapter 10*, the Airflow architecture includes a queue (usually executed by **Celery**) to create an order of execution when we have parallel jobs running simultaneously. The `pool` parameter limits the number of simultaneous jobs. Finally, `priority_weight`, as the name suggests, defines the priority of a DAG over other DAGs.

You can read more about these parameters in the Airflow official documentation here:

`https://airflow.apache.org/docs/apache-airflow/1.10.2/tutorial.html`

See also

You can read more about scheduling with crontab also at `https://crontab.guru/`.

Scheduling historical data ingestion

Historical data is vital for data-driven decisions, providing valuable insights and supporting decision-making processes. It can also refer to data that has been accumulated over a period of time. For example, a sales company can use historical data from previous marketing campaigns to see how they have influenced the sales of a specific product over the years.

This exercise will show how to create a scheduler in Airflow to ingest historical data using the best practices and common concerns related to this process.

Getting ready

Please refer to the *Technical requirements* section for this recipe since we will handle it with the same technology mentioned here.

In this exercise, we will create a simple DAG inside our DAGs folder. The structure of your Airflow folder should look like the following:

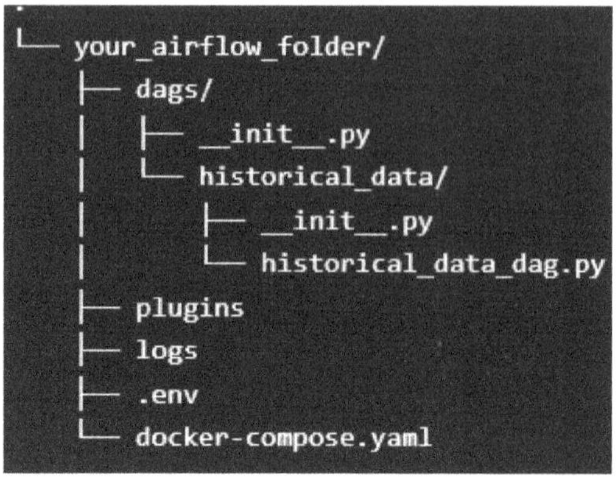

```
└── your_airflow_folder/
    ├── dags/
    │   ├── __init__.py
    │   └── historical_data/
    │       ├── __init__.py
    │       └── historical_data_dag.py
    ├── plugins
    ├── logs
    ├── .env
    └── docker-compose.yaml
```

Figure 11.6 – historical_data_dag folder structure in your local Airflow directory

How to do it...

Here are the steps for this recipe:

1. Let's start by importing our libraries:

```
from airflow import DAG
from airflow.operators.python_operator import PythonOperator

from datetime import datetime, timedelta
```

2. Now, let's define `default_args`. Since we wish to process old data, I will set `datetime` for `start_date` before the current day, and `end_date` will be near the current day. See the following code:

```
default_args = {
    'owner': 'airflow',
    'depends_on_past': False,
    'email': ['airflow@example.com'],
    'email_on_failure': True,
    'email_on_retry': True,
    'retries': 1,
    'retry_delay': timedelta(minutes=5),
    'start_date': datetime(2023, 4, 2),
    'end_date': datetime(2023, 4, 10),
    'schedule_interval': '@daily'
}
```

3. Then, we will create a simple function to print the date Airflow used to execute the pipeline. You can see it here:

```
def my_task(execution_date=None):
    print(f"execution_date:{execution_date}")
```

4. Finally, we will declare our DAG parameters and a `PythonOperator` task to execute it, as you can see here:

```
with DAG(
    'historical_data_dag',
    default_args=default_args,
    description='A simple ETL job using Python commands to
retrieve historical data',
) as dag:

    p1 = PythonOperator(
            task_id="p1",
            python_callable=my_task,
        )
    p1
```

5. Heading to the Airflow UI, let's proceed with the usual steps to enable the DAG and see its execution. On the `historical_data_dag` page, you should see something similar to the following screenshot:

Figure 11.7 – historical_data_dag DAG in the Airflow UI

As you can see, the task ran with success.

6.   Now, let's check our `logs` folder. If we select the folder with the same name as the DAG we created (`historical_data_dag`), we will observe `run_id` instances on different days, beginning on April 2 and finishing on April 10:

Figure 11.8 – Airflow logs folder showing retroactive ingestion

7.   Let's open the first `run_id` folder to explore the log for that run:

Figure 11.9 – DAG log for April 2, 2023

The log tells us the `execution_date` parameter, which is the same as the `start_date` parameter.

Here is a closer look at the logs:

```
[2023-04-12 20:10:25,205] [ ... ] {logging_mixin.py:115} INFO -
execution_date:2023-04-02T00:00:00+00:00
[2023-04-12 20:10:25,205] [ ... ] {python.py:173} INFO - Done.
Returned value was: None
```

We will observe the same pattern for the `run_id` for April 3:

Figure 11.10 – DAG log for April 3, 2023

Here is a closer look at the log output:

```
2023-04-12 20:10:25,276] [ ... ] {logging_mixin.py:115} INFO -
execution_date:2023-04-03T00:00:00+00:00
[2023-04-12 20:10:25,276] [...] {python.py:173} INFO - Done.
Returned value was: None
```

The `execution_date` also refers to April 3.

This shows us that Airflow has used the interval declared on `start_date` and `end_date` to run the task!

Now, let's proceed to understand how the scheduler works.

How it works...

As we saw, scheduling and retrieving historical data with Airflow is straightforward, and the key parameters were `start_date`, `end_date`, and `schedule_interval`. Let's discuss them in a little more detail:

- The `start_date` parameter defines the first date Airflow will look at when the pipeline is triggered. In our case, it was April 2.

- Next is `end_date`. Usually, this is not a mandatory parameter, even for recurrent ingests. However, the purpose of using it was to show that we can set a date as a limit to stop the ingestion.

- Finally, `schedule_interval` dictates the interval between two dates. We used a daily interval in this exercise, but we could also use `crontab` if we needed more granular historical ingestion. We will explore this in more detail in the *Setting up the schedule_interval parameter* recipe.

  With this information, it is easier to understand the logs we got from Airflow:

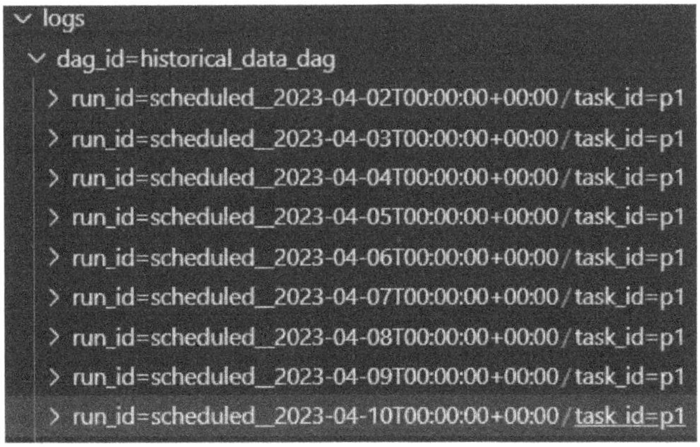

Figure 11.11 – Airflow logs folder showing historic ingestion

Each folder represents one historical ingestion that occurred at a daily interval. Since we did not define a more granular date-time specification, the folder name uses the time that the job was triggered. This information is not included in the logs.

To show what date Airflow was using behind the scenes, we created a simple function:

```
def my_task(execution_date=None):
    print(f"execution_date:{execution_date}")
```

The only purpose of the function is to show the execution date of the task. The `execution_date` parameter is an internal parameter that displays when a task is executed and can be used by operators or other functions to execute something based on a date.

For example, let's say we need to retrieve historical data stored as a partition. We can use `execution_date` to pass the date-time information to a Spark function, which will read and retrieve data from that partition with the same date information.

As you can see, retrieving historical data/information in Airflow requires a few configurations. A good practice is to have a separate and dedicated DAG for historical data processing so that current data ingestion is not impaired. Also, if it is necessary to reprocess data, we can do it with a few parameter changes.

There's more...

Inside the technique of ingesting historical data using Airflow are two important concepts: *catchup* and *backfill*.

Scheduling and running DAGs for past periods that may have been missed for various reasons is commonly known as **catchup** in Apache Airflow. This mechanism allows the system to execute DAGs retrospectively by following their pre-specified `schedule_interval`. By default, this feature is enabled in Airflow. Therefore, if a paused or uncreated DAG's `start_date` lies in the past, it will be automatically scheduled and executed for missed time intervals. The following diagram illustrates this:

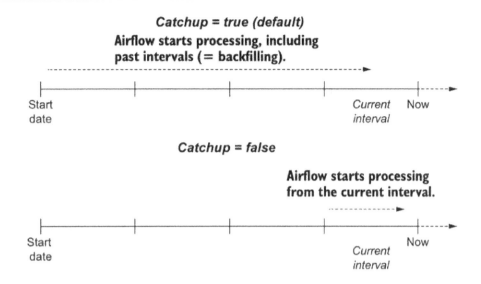

Figure 11.12 – Airflow catchup timeline. Source: https://medium.com/nerd-for-tech/airflow-catchup-backfill-demystified-355def1b6f92

On the other hand, Airflow's *backfill* functionality allows you to execute DAGs retroactively, along with their associated tasks for past periods that may have been missed due to the DAG being paused, not yet created, or for any other reason. Backfilling in Airflow is a powerful feature that helps you to fill the gaps and catch up on data processing or workflow execution that may have been missed in the past.

You can read more about it on *Amit Singh Rathore's* blog page here: `https://medium.com/nerd-for-tech/airflow-catchup-backfill-demystified-355def1b6f92`.

Scheduling data replication

In the first chapter of this book, we covered what data replication is and why it's important. We saw how vital this process is in the prevention of data loss and in promoting recovery from disasters.

Now, it is time to learn how to create an optimized schedule window to make data replication happen. In this recipe, we will create a diagram to help us decide the best moment to replicate our data.

Getting ready

This exercise does not require technical preparation. However, to make it closer to a real scenario, let's imagine we need to decide the best way to ensure the data from a hospital is being adequately replicated.

We will have two pipelines: one holding patient information and another with financial information. The first pipeline collects information from a patient database and synthesizes it into readable reports used by the medical team. The second will feed an internal dashboard used by the hospital executives.

Due to infrastructure limitations, the operations team has only one requirement: only one pipeline can have its data replicated quickly.

How to do it...

Here are the steps for this recipe:

1. **Identify the targets to replicate**: As described in the *Getting ready* section, we have identified the target data, which are the pipeline holding patient information, and the pipeline with financial data to feed a dashboard.

   However, suppose this information is not coming promptly from stakeholders or other relevant people. In that case, we must always start by identifying our project's most critical tables or databases.

2. **Replication periodicity**: We must define the replication schedule based on our data's criticality or relevance. Let's take a look at the following diagram:

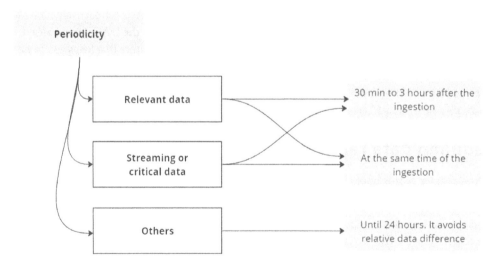

Figure 11.13 – Periodicity of data replication

As we can see, the more critical the data is, the more frequently the replication is recommended. In our scenario, the patient reports would fit better with 30 minutes to 3 hours after the ingestion, while the financial data can be replicated until 24 hours have passed.

3. **Set a schedule window for replication**: Now, we need to create a schedule window to replicate the data. This replication decision needs to take into consideration two important factors, as you can see in the following diagram:

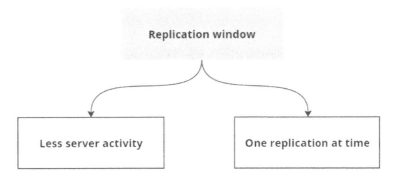

Figure 11.14 – Replication window

Based on the two pipelines (and remembering we need to prioritize one), the suggestion would be to replicate the financial data every day after business working hours, while the patient data can be done at the same time as new information arrives.

Don't worry if this seems a bit confusing. Let's explore the details in the *How it works* section.

How it works...

Data replication is a vital process that ensures data availability and disaster recovery. Its concept is older than the current ETL process and has been used for many years in on-premises databases. Our advantage today is that we can carry out this process at any moment. In contrast, replication had a strict schedule window a few years ago due to hardware limitations.

In our example, we handled two pipelines that had distinct severity levels. The idea behind this is to teach attentive eyes to decide when doing each replication.

The first pipeline, which is the patient reports pipeline, handles sensitive data such as personal information and medical history. It also may be helpful for doctors and other health workers to help a patient.

Based on this, the best approach is to replicate this data within a few minutes or hours of it being processed, allowing high availability and redundancy.

At first look, the financial data seems to be very critical and demands fast replication; we need to remember this pipeline feeds data to a dashboard, and therefore, an analyst can use the raw data to generate reports.

The decision to schedule data replication must consider other factors besides the data involved. It is also essential to understand who is interested in or needs to access the data and how it impacts the project, area, or business when it becomes unavailable.

There's more...

This recipe covered a simple example of setting a scheduling agenda for data replication. We also covered in *step 3* the two main points to have in mind when doing so. Nevertheless, many other factors can influence the scheduler's performance and execution. A few examples are as follows:

- Where Airflow (or a similar platform) is hosted on a server
- The CPU capacity
- The number of schedulers
- Networking throughput

If you want to know more about it, you can find a complete list of these factors in the Airflow documentation: `https://airflow.apache.org/docs/apache-airflow/stable/administration-and-deployment/scheduler.html#what-impacts-scheduler-s-performance`.

The great thing about this documentation is that many points also apply to other data pipeline processors and can serve as a guide.

Setting up the schedule_interval parameter

One of the most widely used parameters in Airflow DAG scheduler configuration is `schedule_interval`. Together with `start_date`, it creates a dynamic and continuous trigger for the pipeline. However, there are some small details we need to pay attention to when setting `schedule_interval`.

This recipe will cover different forms to set up the `schedule_interval` parameter. We will also explore a practical example to see how the scheduling window works in Airflow, making it more straightforward to manage pipeline executions.

Getting ready

While this exercise does not require any technical preparation, it is recommended to take notes about when the pipeline is supposed to start and the interval between each trigger.

How to do it...

Here, we will show only the `default_args` dictionary to avoid code redundancy. However, you can always check out the complete code in the GitHub repository: `https://github.com/PacktPublishing/Data-Ingestion-with-Python-Cookbook/tree/main/Chapter_11/settingup_schedule_interval`.

Let's see how we can declare `schedule_interval`:

- **Using friendly names**: A common way to declare the `schedule_interval` value is by using accessible names such as `@daily`, `@hourly`, or `@weekly`. See what it looks like in the following code:

```
default_args = {
    'owner': 'airflow',
    'depends_on_past': False,
    'retries': 1,
    'retry_delay': timedelta(minutes=5),
    'start_date': datetime(2023, 4, 2),
    'schedule_interval': '@daily'
}
```

- **Using crontab notation:** Airflow also supports defining `schedule_interval` using crontab notation:

```
default_args = {
    'owner': 'airflow',
    'depends_on_past': False,
    'retries': 1,
    'retry_delay': timedelta(minutes=5),
    'start_date': datetime(2023, 4, 2),
    'schedule_interval': '0 22 * * 1-5'
}
```

In this case, we set the scheduler to start every weekday, from Monday to Friday, at 10 pm (or 22:00 hours).

- **Using Python's timedelta function**: Finally, another common way to set `schedule_interval` is by using the `timedelta` method. In the following code, we can set the pipeline to trigger with an interval of one day between each execution:

```
default_args = {
    'owner': 'airflow',
    'depends_on_past': False,
    'retries': 1,
    'retry_delay': timedelta(minutes=5),
```

```
        'start_date': datetime(2023, 4, 2),
        'schedule_interval': timedelta(days=1)
}
```

How it works...

The `schedule_interval` parameter is an essential aspect of scheduling DAGs in Airflow and provides a flexible way to define how frequently your workflows should be executed. We can think of it as the core of the Airflow scheduler.

The goal of this recipe was to show the different ways to set `schedule_interval` and when to use each of them. Let's explore them in more depth:

- **Friendly names:** As the name suggests, this notation uses user-friendly labels or aliases. It provides an easy and convenient way to specify the exact time and date for scheduled tasks to run. It can be an easy and simple solution if you don't have a specific date-time to run the scheduler.

- **Crontab notation:** Crontabs have long been widely used across applications and systems. Crontab notation consists of five fields, representing the minute, hour, day of the month, month, and day of the week. It is a great choice when handling complex schedules, for example, executing the trigger at 1 p.m. on Mondays and Fridays, or other combinations.

- **timedelta function**: This Pythonic technique is commonly used by Airflow users to set the schedule of the DAGs. Using a simple declaration, we can set whether the DAG will run with the interval of one day (as we saw in *step 3*) or every five minutes (`timedelta(minutes=5)`). It is also a user-friendly notation but with more granular power.

> **Note**
>
> Although we have seen three ways to set the `schedule_interval` here, remember Airflow is not a streaming solution, and having multiple DAGs running with a small interval between them can overload the server. Consider using a streaming tool if you or your team needs to schedule ingestions every 10-30 minutes, or less.

See also

TowardsDataScience has a fantastic blog post about how `schedule_interval` works behind the scenes. You can find it here: `https://towardsdatascience.com/airflow-schedule-interval-101-bbdda31cc463`.

Solving scheduling errors

At this point, you may have already experienced some issues with scheduling pipelines not being triggered as expected. If not, don't worry; it will happen sometime and is totally normal. With several pipelines running in parallel, in different windows, or attached to different timezones, it is expected to be entangled with one or another.

To avoid this entanglement, in this exercise, we will create a diagram to assist in the debugging process, identify the possible causes of a scheduler not working correctly in Airflow, and see how to solve it.

Getting ready

This recipe does not require any technical preparation. Nevertheless, taking notes and writing down the steps we will follow here can be helpful. Writing when learning something new can help to fix the knowledge in our minds, making it easier to remember later.

Back to our exercise; scheduler errors in Airflow typically give the DAG status None, as shown here:

Figure 11.15 – DAG in the Airflow UI with an error in the scheduler

We will now find out how to fix the error and make the job run again.

How to do it...

Let's try to identify what could be the cause of the error in our scheduler. Don't worry about understanding why we used the approaches that we have. We will cover them in detail in *How it works*:

1. We can first check whether start_date has been set to datetime.now(). If this is the case, the best approach here is to change this parameter value to a specific date, as you can see here:

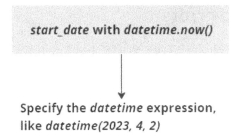

Figure 11.16 – Error caused by start_date parameter

The code will look like this:

```
default_args = {
    'owner': 'airflow',
    'depends_on_past': False,
    'retries': 1,
    'retry_delay': timedelta(minutes=5),
    'start_date': datetime(2023, 4, 2),
    'schedule_interval': '@daily'
}
```

2. Now we can verify whether `schedule_interval` is aligned with the `start_date` parameter. In the following diagram, you can see three possibilities to fix the issue:

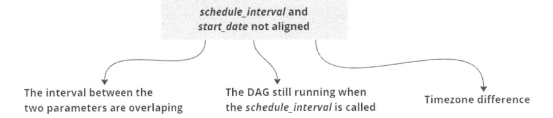

Figure 11.17 – Error caused by the start_date and schedule_interval parameters

You can prevent this error by using crontab notation in `schedule_interval`, as follows:

```
schedule_interval='0 2 * * *'
```

If you are facing problems with the timezone, you can define which timezone Airflow will trigger the job in by using the pendulum library:

```
pendulum.now("Europe/Paris")
```

3. Finally, another standard error scenario is when schedule_interval changes after the DAG has been running for some time. In this case, the solution usually is to recreate the DAG:

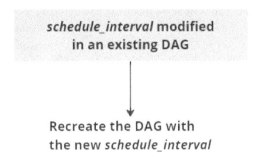

Figure 11.18 – Error caused by a change in schedule_interval

At the end of these steps, we will end up with a debug diagram similar to this:

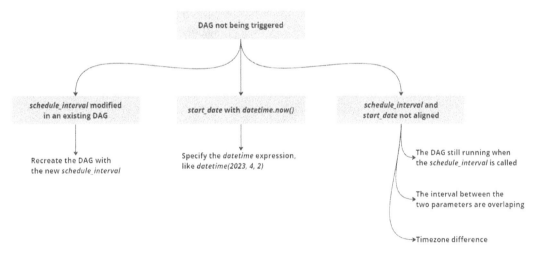

Figure 11.19 – Diagram to assist in identifying an error caused in the Airflow scheduler

How it works...

As you can see, the goal of this recipe was to show three different scenarios in which it is common to observe errors related to the scheduler. Errors in the scheduler normally lead to a DAG status as None, as we saw in the *Getting ready* section. However, having a trigger that does not behave as expected is also considered an error. Now, let's explore the three addressed scenarios and their solutions.

The first scenario usually occurs when we want to use the current date for start_date. Although it seems like a good idea to use the datetime.now() function to represent the current date-time, Airflow will not interpret it as we do. The datetime.now() function will create what we call a *dynamic scheduler*, and the trigger will never be executed. It happens because the execution schedule uses start_date and schedule_interval to know when to execute the trigger, as you can see here:

$$\text{start\_date} + \text{schedule\_interval} = \text{execution schedule}$$

Figure 11.20 – Airflow execution scheduler equation

If we use datetime.now(), it moves along with time and will never be triggered. We recommend using a static schedule definition, as we saw in *step 1*.

A typical error is when start_date and schedule_interval are not aligned. Based on the explanation of *Figure 11.20*, we can already imagine why aligning these two parameters and preventing overlapping are so important. As addressed in *step 2*, a good way to prevent this is by using crontab notation to set schedule_interval.

A vital topic is the timezones involved in the process. If you look closely at the top of the Airflow UI, you will see a clock and its associated timezone, as follows:

Figure 11.21 – Airflow UI clock with the timezone displayed

This indicates that the Airflow server is running in the UTC timezone, and all DAGs and tasks will be executed using the same logic. If you are working in a different timezone and want to ensure it will run according to your timezone, you can use the pendulum library, as you can see here:

```
schedule_interval = pendulum.now("Europe/Paris")
```

pendulum is a third-party Python library that provides easy date-time manipulations using the built-in datetime Python package. You can find out more about it in the pendulum official documentation: https://pendulum.eustace.io/.

Finally, the last scenario has a straightforward solution: recreate the DAG if `schedule_interval` changes after some executions. Although this error may not always occur, it is a good practice to recreate the DAG to prevent further problems.

There's more...

We have provided in this recipe some examples of what you can check if the scheduler is not working, but other common errors in Airflow can happen. You can find out more about this on *Astronomer's* blog page here: `https://www.astronomer.io/blog/7-common-errors-to-check-when-debugging-airflow-dag/`.

In the blog, you can find other scenarios where Airflow throws a silent error (or an error without an explicit error message) and how to solve them.

Further reading

- `https://airflow.apache.org/docs/apache-airflow/stable/faq.html#what-s-the-deal-with-start-date`

- `https://se.devoteam.com/expert-view/why-my-scheduled-dag-does-not-runapache-airflow-dynamic-start-date-for-equally-unequally-spaced-interval/`

- `https://stackoverflow.com/questions/66098050/airflow-dag-not-triggered-at-schedule-time`

Using Data Observability for Debugging, Error Handling, and Preventing Downtime

We are reaching the end of our journey through the data ingestion world and have covered many important topics and seen how they could be applied to real-life projects. Now, to finish this book with a flourish, the final topic is the concept of **data observability**.

Data observability refers to the ability to monitor, understand, and troubleshoot the health, quality, and other vital aspects of data in a big organization or a small project. In summary, it ensures that data is accurate, reliable, and available when needed.

Although each recipe in this chapter can be executed separately, the goal is to configure tools that, when set together, create a monitoring and observability architecture ready to bring value to a project or team.

You will learn about the following recipes:

- Setting up StatsD for monitoring
- Setting up Prometheus for storing metrics
- Setting up Grafana for monitoring
- Creating an observability dashboard
- Setting custom alerts or notifications

Technical requirements

This chapter requires that Airflow is installed on your local machine. You can install it directly on your **Operating System (OS)** or using a Docker image. For more information, refer to *Chapter 1*, and the *Configuring Docker for Airflow* recipe.

After following the steps described in *Chapter 1*, ensure Airflow runs correctly. You can do that by checking the Airflow UI at this link: `http://localhost:8080`

If you are using a Docker container (as I am) to host your Airflow application, you can check its status in the terminal by using the following command:

```
$ docker ps
```

Here is the status of the container:

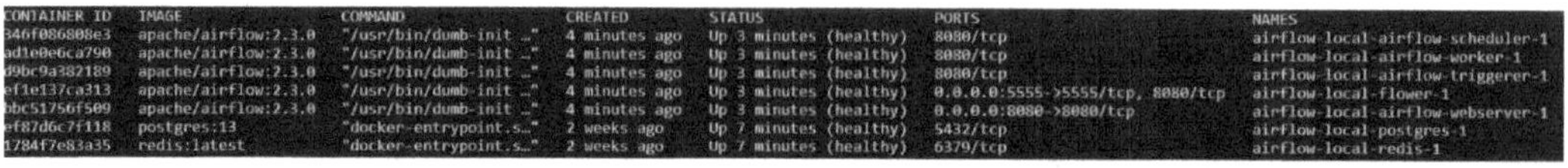

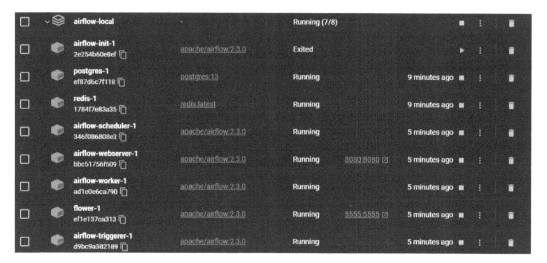

Figure 12.1 – Airflow containers running

Figure 12.2 – Docker Desktop view of Airflow containers running

Docker images

This chapter requires the creation of other Docker containers to build the monitoring and observability architecture. If you are using `docker-compose.yaml` file to run your Airflow application, you can add the other images addressed here to the same `docker-compose.yaml` file and run it all together.

If you are running Airflow locally, you can create and configure each Docker image separately or create a `docker-compose.yaml` file just for the monitoring tools approach in this chapter.

Setting up StatsD for monitoring

As introduced in *Chapter 10*, **StatsD** is an open source daemon that gathers and aggregates metrics about application behaviors. Due to its flexibility and lightweight, StatsD is used on several monitoring and observability tools, such as **Grafana**, **Prometheus**, and **ElasticSearch**, to visualize and analyze the collected metrics.

In this recipe, we will configure StatsD using a Docker image as the first step in building a monitoring pipeline. Here, StatsD will collect and aggregate Airflow information and make it available to Prometheus, our monitoring database, in the *Setting up Prometheus for storing metrics* recipe.

Getting ready

Refer to the *Technical requirements* section for this recipe since we will handle it with the same technology.

How to do it...

Here are the steps to perform this recipe:

1. Let's start by defining our Docker configurations for StatsD. These lines will be added under the `services` section inside the `docker-compose` file:

    ```
    statsd-exporter:
      image: prom/statsd-exporter
      container_name: statsd-exporter
      command: "--statsd.listen-udp=:8125 --web.listen-
    address=:9102"
      ports:
        - 9102:9102
        - 8125:8125/udp
    ```

2. Next, let's set the Airflow environment variables to install StatsD and export the metrics to it, as you can see here:

    ```
    # StatsD configuration
    AIRFLOW__SCHEDULER__STATSD_ON: 'true'
    AIRFLOW__SCHEDULER__STATSD_HOST: statsd-exporter
    AIRFLOW__SCHEDULER__STATSD_PORT: 8125
    AIRFLOW__SCHEDULER__STATSD_PREFIX: airflow
    _PIP_ADDITIONAL_REQUIREMENTS: ${_PIP_ADDITIONAL_REQUIREMENTS:-
    apache-airflow[statsd]}
    ```

    If you need help to set these variables in Airflow, please refer to *Chapter 10*, and the *Configuring logs in airflow.cfg* recipe.

Your Airflow variables in the `docker-compose` file should look like this:

```
version: '3'
x-airflow-common:
  &airflow-common
  # In order to add custom dependencies or upgrade provider packages you can use your extended image.
  # Comment the image line, place your Dockerfile in the directory where you placed the docker-compose.yaml
  # and uncomment the "build" line below, Then run `docker-compose build` to build the images.
  image: ${AIRFLOW_IMAGE_NAME:-apache/airflow:2.3.0}
  # build: .
  environment:
    &airflow-common-env
    AIRFLOW__CORE__EXECUTOR: CeleryExecutor
    AIRFLOW__DATABASE__SQL_ALCHEMY_CONN: postgresql+psycopg2://airflow:airflow@postgres/airflow
    # For backward compatibility, with Airflow <2.3
    AIRFLOW__CORE__SQL_ALCHEMY_CONN: postgresql+psycopg2://airflow:airflow@postgres/airflow
    AIRFLOW__CELERY__RESULT_BACKEND: db+postgresql://airflow:airflow@postgres/airflow
    AIRFLOW__CELERY__BROKER_URL: redis://:@redis:6379/0
    AIRFLOW__CORE__FERNET_KEY: ''
    AIRFLOW__CORE__DAGS_ARE_PAUSED_AT_CREATION: 'true'
    AIRFLOW__CORE__LOAD_EXAMPLES: 'true'
    AIRFLOW__API__AUTH_BACKENDS: 'airflow.api.auth.backend.basic_auth'
    _PIP_ADDITIONAL_REQUIREMENTS: ${_PIP_ADDITIONAL_REQUIREMENTS:-apache-airflow-providers-mongo apache-airflow[statsd]}
    # StatsD configuration
    AIRFLOW__SCHEDULER__STATSD_ON: 'true'
    AIRFLOW__SCHEDULER__STATSD_HOST: statsd-exporter
    AIRFLOW__SCHEDULER__STATSD_PORT: 8125
    AIRFLOW__SCHEDULER__STATSD_PREFIX: airflow
```

Figure 12.3 – Airflow environment variables with StatsD configurations

3. Now, restart your Docker containers to apply the configurations.

4. Once you do so, and all containers are up and running, let's check the `http://localhost:9102/` address in a browser. You should see the following page:

StatsD Exporter

Metrics

Figure 12.4 – StatsD page in the browser

5. Then, click on **Metrics**, and a new page will appear showing something similar to the following:

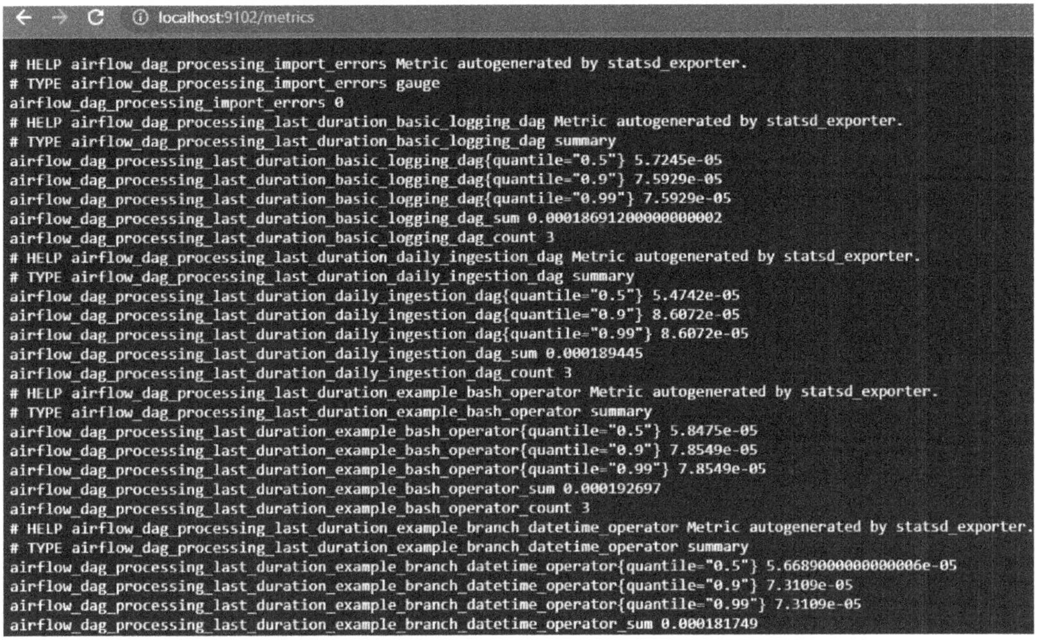

Figure 12.5 – StatsD metrics being shown in the browser

The lines shown in the browser confirm StatsD is successfully installed and collecting data from Airflow.

How it works...

As you can observe, configuring StatsD with Airflow is very straightforward. In fact, StatsD is not new for us since we already covered it in *Chapter 10*, in the *Designing advanced monitoring* recipe. However, let's recap some of the concepts.

StatsD is an open source daemon tool built by Etsy employees that receives information via the **User Datagram Protocol** (**UDP**), making it fast and lightweight since it discards the necessity of sending a confirmation message back to the sender.

Now, looking at the code, the first thing we did was to set the Docker container to run StatsD. Alongside all the usual parameters to run a container, the key point is the command parameter, as follows:

```
command: "--statsd.listen-udp=:8125 --web.listen-address=:9102"

# StatsD configuration
AIRFLOW__SCHEDULER__STATSD_ON: 'true'
AIRFLOW__SCHEDULER__STATSD_HOST: statsd-exporter
AIRFLOW__SCHEDULER__STATSD_PORT: 8125
```

```
AIRFLOW__SCHEDULER__STATSD_PREFIX: airflow
_PIP_ADDITIONAL_REQUIREMENTS: ${_PIP_ADDITIONAL_REQUIREMENTS:-apache-
airflow[statsd]}
```

See also

You can check the Docker image of StatsD on the **Docker Hub** page here: `https://hub.docker.com/r/prom/statsd-exporter`

Setting up Prometheus for storing metrics

Although it is generally called a database, Prometheus is not a traditional database like MySQL. Instead, its structure is more similar to a time-series database designed for monitoring and observability purposes.

Due to its flexibility and power, this tool is widely used by DevOps and **Site Reliability Engineers (SREs)** to store metrics and other relevant information about systems and applications. Together with Grafana (which we will explore in later recipes), it is one of the most used monitoring tools in projects and by teams.

This recipe will configure a Docker image to run a Prometheus application. We will also connect it to StatsD to store all the metrics generated.

Getting ready

Refer to the *Technical requirements* section for this recipe since we will handle it with the same technology.

How to do it...

Here are the steps to perform this recipe:

1. Let's begin by adding the following lines to our `docker-compose` file under the `services` section:

```
prometheus:
  image: prom/prometheus
  ports:
  - 9090:9090
  links:
    - statsd-exporter # Use the same name as your statsd
container
  volumes:
    - ./prometheus:/etc/prometheus
  command:
    - '--config.file=/etc/prometheus/prometheus.yml'
    - --log.level=debug
```

```
      - --web.listen-address=:9090
      - --web.page-title='Prometheus - Airflow Metrics'
```

2. Now, create a folder named prometheus at the same level as your docker-compose file. Inside the folder, create a new file named prometheus.yml with the following code and save it:

```
scrape_configs:
  - job_name: 'prometheus'
    static_configs:
      - targets: ['prometheus:9090']
  - job_name: 'statsd-exporter'
    static_configs:
      - targets: ['statsd-exporter:9102']
```

On static_configs, make sure the target has the same name and the exposed port of the StatsD container. Otherwise, you will face problems in establishing a connection with the container.

3. Now, restart your Docker containers.

4. When the containers are back up and running, access the following link in your browser: http://localhost:9090/.

    You should see a page like the following:

Figure 12.6 – Prometheus UI

5.   Now, click on the list icon next to the **Execute** button on the right of the page. It will open a list with all metrics available to be used. If everything is well configured, you should see something like the following:

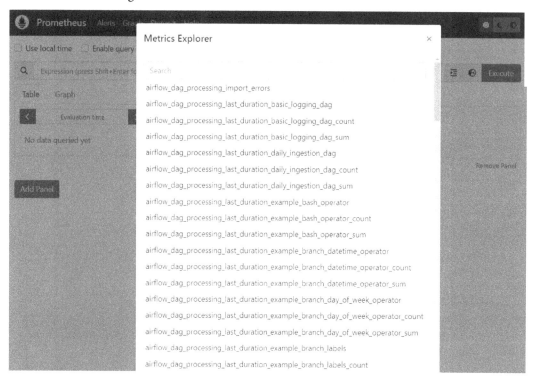

Figure 12.7 – Prometheus available metric list

We have successfully set up Prometheus, which is already storing the metrics sent by StatsD!

How it works...

Let's explore in more depth what we did in this exercise by examining the container definitions in *Step 1*. Since we already have basic knowledge of Docker, we will cover the most critical parts of the container settings.

The first thing that draws attention is the links section in the docker-compose file. In this section, we declared that the Prometheus container must be connected and linked to the StatsD container configured in the *Setting up StatsD for monitoring* recipe:

```
links:
    - statsd-exporter # Use the same name as your statsd container
```

Next, we set `volumes` to reflect a local folder to a folder inside the container. This step is essential because then we can also mirror the configuration file of Prometheus:

```
volumes:
  - ./prometheus:/etc/prometheus
```

Finally, in the `command` section, we declared where the configuration file will be placed inside the container and other minor settings:

```
command:
  - '--config.file=/etc/prometheus/prometheus.yml'
  - --log.level=debug
  - --web.listen-address=:9090
  - --web.page-title='Prometheus - Airflow Metrics'
```

Then, the following steps were dedicated to setting the Prometheus configuration file, as you can see here:

```
scrape_configs:
  - job_name: 'prometheus'
    static_configs:
      - targets: ['prometheus:9090']
  - job_name: 'statsd-exporter'
    static_configs:
      - targets: ['statsd-exporter:9102']
```

By definition, Prometheus collects metrics from itself and other applications through an HTTP request. In other words, it parses the response and ingests the collected samples for storage. That's why we used `scrape_configs`.

If you look closely, you will observe that we declared two scrape jobs: one for Prometheus and another for StatsD. Due to that configuration, we could see Airflow metrics in the Prometheus metrics list. If we needed to include any other scrape configuration, we would just need to edit the local `prometheus.yml` file and restart the server.

Many other configurations are available in Prometheus, such as setting the scrape interval. You can read more about its configurations on the official documentation page at `https://prometheus.io/docs/prometheus/latest/getting_started/`.

There's more...

In this recipe, we saw how to set Prometheus to store metrics coming from StatsD. This time-series database also has other capabilities, such as creating small visualizations in the web UI and connecting with other client libraries, and has an alerting system called Alertmanager.

If you want to go deeper into how Prometheus works and other functionalities, Sudip Sengupta has a fantastic blog post about it, which you can read here:

`https://www.airplane.dev/blog/prometheus-metrics`

Setting up Grafana for monitoring

Grafana is an open source tool built to create visualizations and monitor data from other systems and applications. Together with Prometheus, it is one of the most popular DevOps tools due to its flexibility and rich features.

In this exercise, we will configure a Docker image to run Grafana and connect it to Prometheus. This configuration will not only give us the ability to explore the Airflow metrics even further but also the opportunity to learn in practice how to work with a set of the most popular tools for monitoring and observability.

Getting ready

Refer to the *Technical requirements* section for this recipe since we will handle it with the same technology.

In this recipe, I will use the same `docker-compose.yaml` file of Airflow and will keep the configurations from the *Setting up StatsD for monitoring* and *Setting up Prometheus for storing metrics* recipes, to connect them and proceed with the monitoring and observability architecture.

How to do it...

Perform the following steps to try this recipe:

1. As shown in the following, let's add the Grafana container information to our `docker-compose` file as usual. Make sure it is under the `services` section:

    ```
    grafana:
      image: grafana/grafana:latest
      container_name: grafana
      environment:
        GF_SECURITY_ADMIN_USER: admin
        GF_SECURITY_ADMIN_PASSWORD: admin
        GF_PATHS_PROVISIONING: /grafana/provisioning
      links:
        - prometheus # use the same name of your Prometheus docker
    container
      ports:
    ```

```
    - 3000:3000
volumes:
    - ./grafana/provisioning:/grafana/provisioning
```

Feel free to use a different administrator username as a password.

2. Now, create a folder called `grafana` on the same level as your Docker file, and restart your containers.

3. After it is back up and running, insert the `http://localhost:3000/login` link in your browser. A login page similar to this will appear:

Figure 12.8 – Grafana login page

It confirms Grafana is set up correctly!

4. Then, let's use the administrator credentials to log in to the Grafana dashboard. After authenticating, you should see the main page as follows:

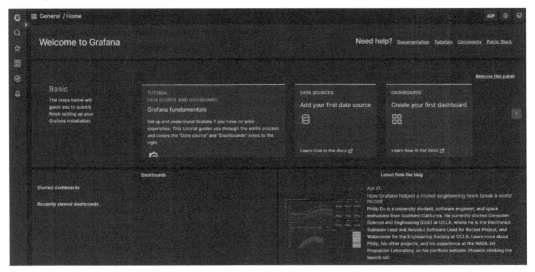

Figure 12.9 – Grafana main page

Since it is our first login, this page has nothing to show. We will take care of visualizations in the *Creating an observability dashboard* recipe.

5. Now, let's add Prometheus as a data source to Grafana. On the bottom-left side of the page, hover your cursor over the engine icon. On the **Configuration** menu, select **Data sources**. See the following screenshot for reference:

Figure 12.10 – Grafana Configuration menu

6.   On the **Data Sources** page, select the Prometheus icon. You will be redirected to a new page showing fields to insert Prometheus settings, as you can see here:

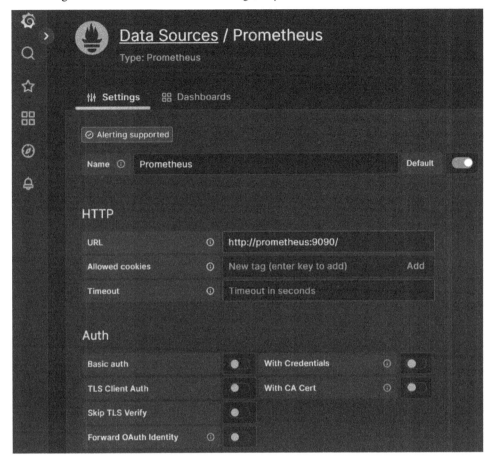

Figure 12.11 – Data Sources page in Grafana

Insert a name for this data source. In the **URL** field, under **HTTP**, insert the link `http://prometheus:9090`. Make sure it has the same name as your Docker container for Prometheus.

Save this configuration, and we have successfully configured Grafana with Prometheus!

How it works...

In this exercise, we saw how simple it is to configure Grafana and integrate it with Prometheus as a data source. In fact, almost all Grafana integrations are very straightforward, requiring just a few pieces of information.

Let's now explore some of our Grafana container settings. Despite the standard Docker container settings, a few items require attention, as you can see here:

```
grafana:
  ...
  environment:
    GF_SECURITY_ADMIN_USER: admin
    GF_SECURITY_ADMIN_PASSWORD: admin
    GF_PATHS_PROVISIONING: /grafana/provisioning
  ...
  volumes:
    - ./grafana/provisioning:/grafana/provisioning
```

The first things are the `environment` variables, where we define the administrator credentials that allow the first login. Then, we declared the path of Grafana provisioning, and, as you will have noticed, we also inserted this path in the `volumes` section.

It is inside the `provisioning` folder where we will have configuration files for data sources connections, plugins, dashboards, and much more. A configuration like this allows more reliability and version control of dashboards and panels. We could also create the Prometheus data source connection using a .yaml configuration file and place it under the `provisioning` and `datasources` folder. It would look similar to this:

```
apiVersion: 1
datasources:
  - name: Prometheus
    type: prometheus
    access: proxy
    url: http://prometheus:9090
```

Any additional data sources can be placed inside this YAML file. You can explore more about the provisioning configurations in Grafana on the official documentation page at `https://grafana.com/docs/grafana/latest/administration/provisioning/`.

With this, we created a simple and efficient monitoring and observability architecture capable of collecting metrics from Airflow (or any other application if needed), storing, and showing them. The architecture can be defined as follows:

Figure 12.12 – Monitoring and observability high-level architecture

We can now start creating our first dashboard and alerts in the two final recipes of this chapter!

There's more...

Besides Prometheus, Grafana has built-in core data source integrations for many applications. It allows easy configuration and a quick setup, which brings a lot of value and maturity to a project. You can find more here: `https://grafana.com/docs/grafana/latest/datasources/#built-in-core-data-sources`.

Grafana Cloud

Grafana Labs has also made the platform available as fully managed and deployed on the cloud. It is a great solution for teams that don't have a dedicated operations team to support and maintain Grafana. Find more information here: `https://grafana.com/products/cloud/`.

Creating an observability dashboard

Now, with our tools up and running, we can finally jump into the visualization dashboards. Monitoring and observability dashboards are designed to help gain deep insights into the health and behavior of our systems. You will observe in this exercise how Grafana can help us create an observability dashboard and a number of features inside it.

In this recipe, we will create our first dashboard with a few panels to better monitor our Airflow application. You will notice that, with a few steps, it is possible to have an overview of how Airflow behaves over time and be prepared to build your future panels.

Getting ready

Refer to the *Technical requirements* section for this recipe since we will handle it with the same technology.

To accomplish this exercise, ensure that StatsD, Prometheus, and Grafana are adequately configured and running.

How to do it...

Let's create our dashboard to keep track of Airflow:

1. On the Grafana main page, hover the cursor over the four-squares icon on the left side panel. Then, select **New dashboard**, as you can see in the following screenshot:

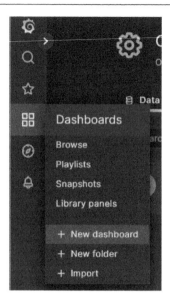

Figure 12.13 – Grafana Dashboards menu

If you need help accessing Grafana, refer to the *Setting up Grafana for monitoring* recipe.

2. You will be redirected to an empty page with the title **New dashboard**. At the top right of the page, select **Save**, insert the name of your dashboard, and click the **Save** button again. Refer to the following screenshot:

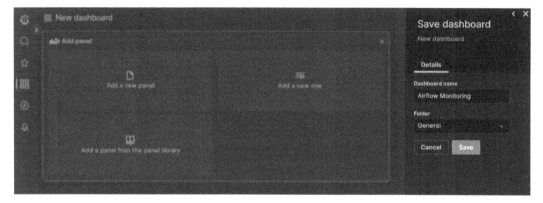

Figure 12.14 – New dashboard page

3. Now, let's create our first panel by clicking on the **Add panel** icon at the top right of the dashboard page, as you can see in the following screenshot:

Figure 12.15 – Add panel icon

4. Now, let's create a panel to show the number of DAGs inside Airflow. On the **Edit Panel** page, set the following information:

    - **Metric**: **airflow_dagbag_size**

    - **Label filters**: **job**, **statsd-exporter**

    - Visualization type: **Stat**

    You can see the filled information in the following screenshot:

Figure 12.16 – Airflow number of DAGs panel count

Click on **Apply** to save and return to the dashboard page.

5. Let's do the same as *Step 3* to create another panel. This time we will create a panel to show the number of Airflow import errors. Fill the fields with the following values:

    • **Metric**: **airflow_dag_processing_import_errors**

    • **Label filters**: **job, statsd-exporter**

    • Visualization type: **Stat**

    You can see the added information in the following screenshot:

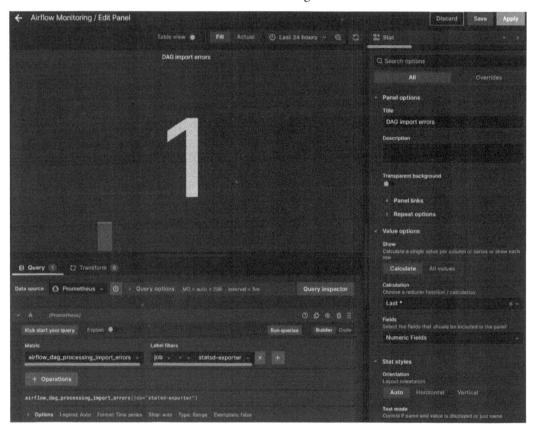

Figure 12.17 – DAG import errors panel count

6. Now, let's create two more panels with the following information:

    • **Metric**: `airflow_executor_queued_tasks`

    • **Label filters**: `job`, `statsd-exporter`

    • Visualization type: **Stat**

    • **Metric**: **airflow_scheduler_tasks_running**

- Label filters: job, statsd-exporter
- Visualization type: **Stat**

7. Let's create two more panels to show the execution time for two different DAGs. Create two panels with the following values:

- Metric: **airflow_dag_processing_last_duration_basic_logging_dag**
- Label filters: **quantile**, **0.99**
- Visualization type: **Time-series**

Refer to the following screenshot:

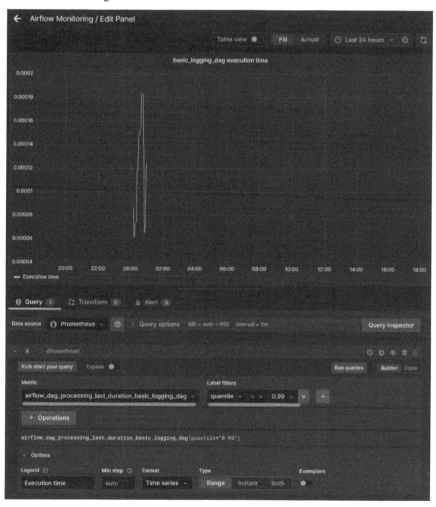

Figure 12.18 – basic_logging_dag execution run panel

- **Metric**: **airflow_dag_processing_last_duration_holiday_ingest_dag**

- **Label filters**: **quantile**, **0.99**

- Visualization type: **Time-series**

    You can see the completed fields in the following screenshot:

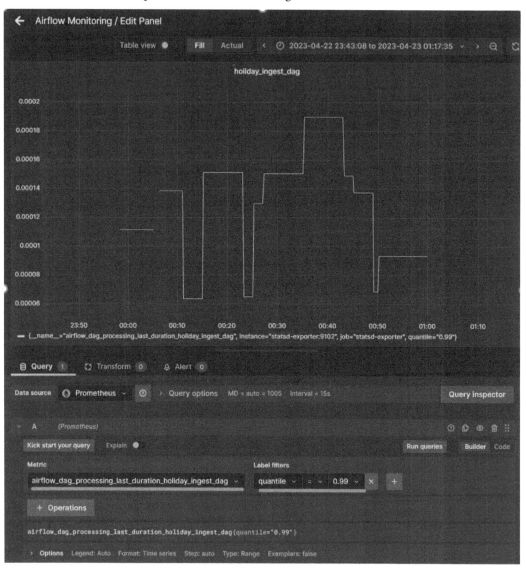

Figure 12.19 – holiday_ingest_dag execution run panel

In the end, you will end up with a dashboard similar to the following:

Figure 12.20 – Complete Airflow Monitoring dashboard view

Don't worry if your dashboard layout does not look exactly like *Figure 12.20*. You can rearrange the panel as much as you want to add your own touch!

How it works…

There are many DevOps visualization tools available on the market. However, most require a paid subscription or trained people to build the panels. As you can observe in this exercise, creating the first dashboard and panels using Grafana can be pretty simple. Of course, as you practice and study advanced concepts in Grafana, you will observe many opportunities to improve and enhance your dashboard.

Now, let's explore the six panels we have created. The idea behind these panels was to create a small dashboard with a minimum of information that could already bring value.

The first four panels give quick and relevant information about Airflow, as follows:

Figure 12.21 –Airflow Monitoring counter panels

They show information about the number of DAGs, how many import errors we have, the number of tasks waiting to be executed, and how many are being executed, respectively. Even though it seems simple, these pieces of information give an overview (therefore, observability) of Airflow's current behavior.

The last two panels show information about the duration of two DAG executions, as follows:

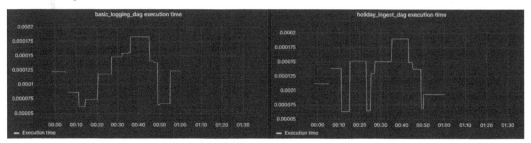

Figure 12.22 – Airflow Monitoring time-series panels

Knowing how much time a DAG takes to run is vital information, and it can offer insight to improve the code or check whether the data used in the pipeline is reliable. For example, if the DAG executes all tasks in less than half the expected time, it can be a sign no data was processed correctly.

Lastly, you can create more dashboards and organize them into folders according to the subject. You can check the recommended best practices for dashboard organization in Grafana's official documentation here: `https://grafana.com/docs/grafana/latest/dashboards/build-dashboards/best-practices/`.

There's more...

Unfortunately, since we have limited data to show on a dashboard, this exercise might not be as fancy as you expected. However, you can explore Grafana panel configurations and master them for further projects using the Grafana playground here: `https://play.grafana.org/d/000000012/grafana-play-home?orgId=1`.

On the **Grafana Play Home** page, you will be able to see different types of panel applications and explore how they were built.

Setting custom alerts or notifications

After configuring our first dashboard to be aware of the Airflow application, we must ensure our monitoring is never left without observation. With teams busy with other tasks, creating alerts is the best way to guarantee we still have oversight over the application.

There are many ways to create alerts and notifications, and previously we implemented something similar to monitor our DAG by sending an email notification when an error occurs. Now, we will try a different approach, using an integration with **Telegram**.

In this recipe, we will integrate Grafana alerts with Telegram. Using a different tool to provide system alerts can help us understand the best approach to advise our teams and break the cycle of always using email.

Getting ready

Refer to the *Technical requirements* section for this recipe since we will handle it with the same technology.

To accomplish this exercise, ensure that StatsD, Prometheus, and Grafana are adequately configured and running. It is also required to have a Telegram account for this exercise. You can find the steps to create an account here: `https://www.businessinsider.com/guides/tech/how-to-make-a-telegram-account`.

How to do it...

Here are the steps to perform this recipe:

1. Let's start by creating a bot on Telegram to be used by Grafana to send the alerts. On the Telegram main page, search for `@BotFather` and start a conversation as follows:

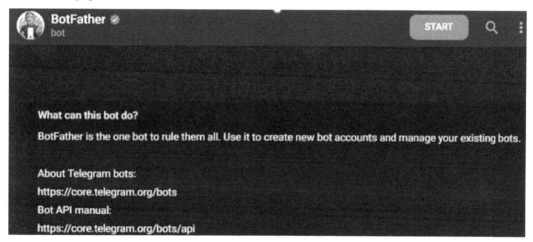

Figure 12.23 – Telegram BotFather

2. Then, type `/newbot` and follow the prompt instructions. BotFather will send you a bot token. Please keep it in a safe place; we will use it later. The message looks like the following:

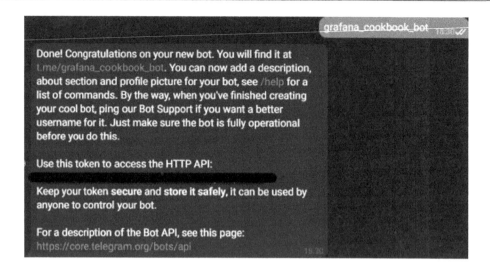

Figure 12.24 – New bot creation message

3. Next, create a group on Telegram and invite your bot to it with administrator privileges.

4. Now, let's use the Telegram API to check the channel ID where the bot is. You can do it by using the following address in your browser:

```
https://api.telegram.org/bot<YOUR CODE HERE>/getUpdates
```

You should see a similar output in the browser:

{"ok":true,"result":[{"update_id":380740357,
"message":{"message_id":3,"from":{"id":154783691,"is_bot":false,"first_name":"Glaucia","last_name":"Esppenchutz","username":   ,"language_code":"en"},"chat":
{"id":154783691,"first_name":"Glaucia","last_name":"Esppenchutz","username":   ,"type":"private"},"date":1682272386   :"/start","entities":
[{"offset":0,"length":6,"type":"bot_command"}]}}]}

Figure 12.25 – Telegram API message with Chat ID

We will use the `id` value later, so keep this in a safe place too.

5. Then, let's proceed to create a Grafana notification group. On the left menu bar, hover your cursor over the bell icon, and select **Contact points**, shown as follows:

Figure 12.26 – Grafana Alerting menu

6.   On the **Contact points** tab, select **Add contact point** as follows:

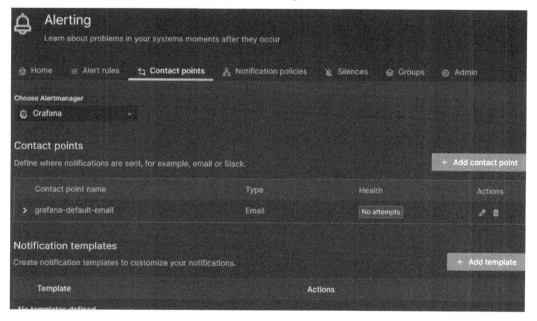

Figure 12.27 – Contact points tab in Grafana

7. Add a name on the **New contact point** page and choose **Telegram** in the **Integration** drop-down menu. Then, complete the **Bot API Token** and **Chat ID** fields. You can see what it looks like here:

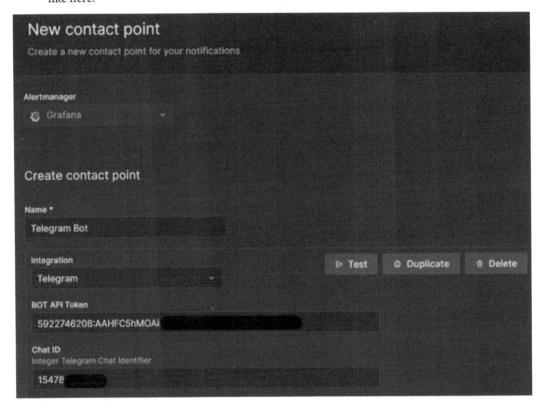

Figure 12.28 – New contact point page

8. Now, let's ensure we inserted the values correctly while selecting the **Test** button. If everything is well configured, you will receive a message on the channel you have your bot in, as follows:

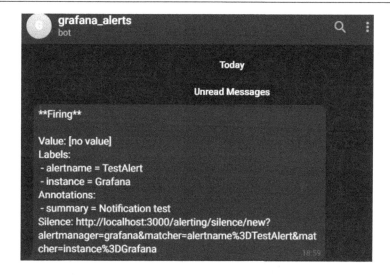

Figure 12.29 – Grafana test message working successfully

It means our bot is ready! Save the contact point and go back to the alerts page.

9. In **Notification policies**, edit the **Root policy** contact point to your Telegram bot as follows:

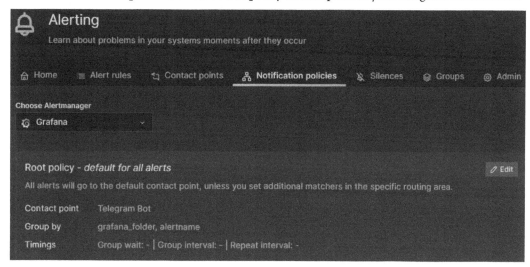

Figure 12.30 – Grafana Notification policies tab

10. Finally, let's create an alert rule to trigger an alert notification. On the **Alert rules** page, select **Create alert rule** to be redirected to a new page. Insert the following values in the fields on this page:

- **Rule name**: Import errors

- **Metric**: airflow_dag_processing_import_errors

- **Label filters: instance, statsd-exporter:9102**
- **Threshold: Input A, IS ABOVE 1**
- **Folder**: Create a new folder called **Errors** and **test_group** in **Evaluation group**
- **Rule group evaluation interval: 3 minutes**

You should have something similar to the following screenshot. You can also use it as a reference to fill in the fields:

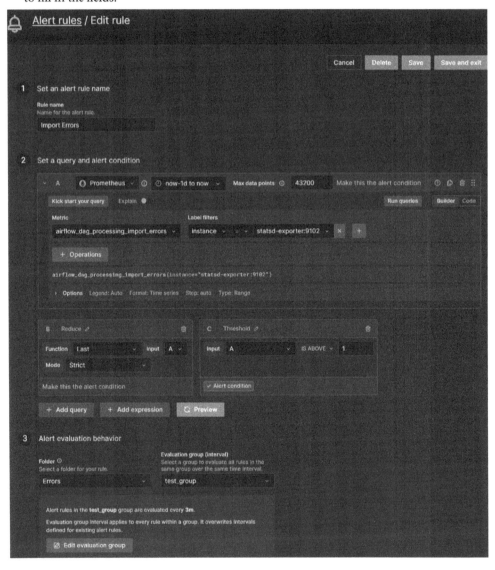

Figure 12.31 – New alert rule for Airflow import errors on Grafana

Save it, and let's simulate an import error in Airflow.

11. After creating any import error in a DAG on Airflow, you will receive a notification on the Telegram channel similar to the following:

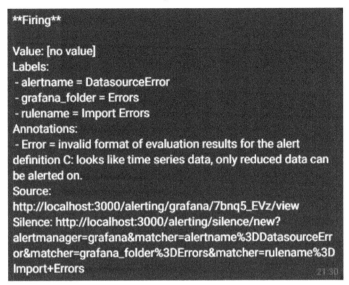

Figure 12.32 – Telegram bot showing a notification after being triggered by a Grafana alert

Since this is a local test, you don't need to worry about the `Annotations` part for now.

Our Grafana notification works, and it is fully integrated with Telegram!

How it works...

Although this recipe has many steps, the content is not complex. This exercise aims to give you a practical end-to-end example of configuring a simple bot to create alerts whenever needed.

Bots are frequently used in DevOps as a tool for notifications of an action, and it was no different here. From *Step 1* to *Step 4*, we focused on configuring a bot in Telegram and a channel where Grafana notifications could be sent. There is no particular reason for choosing Telegram as our messenger, other than the ease of creating an account. Usually, messengers such as **Slack** or **Microsoft Teams** are the favorites of operation teams, and plenty of online tutorials show how to use them.

After configuring the bot, we proceeded to connect it with Grafana. The configuration only required a few pieces of information, such as an authentication token (to control the bot) and the channel's ID. As you observed, many types of integrations are available, and more can be added when installing a plugin. You can see the complete list of plugins here: `https://grafana.com/grafana/plugins/`.

If we needed more than one contact point, we could create it on the **Contact points** tab and create a notification policy to include the new contact as a point to be notified.

Finally, we created an alert rule based on the number of Airflow import errors. Import errors can impair the execution of one or more DAGs; therefore, they are relevant items to monitor.

There are two ways to create an alert and notification: on the **Alert rules** page and directly on a dashboard panel. The latter depends on the panel type, and not all of the panels support integrated alerts. The safest option, and the best practice, is to create an alert rule on the **Alert rules** page.

Creating an alert is similar to a panel, where we need to identify metrics and labels, and the critical points are the **Threshold** and **Alert Evaluation** conditions. These two configurations will determine the limit of metric value acceptance and how long it can take. We set a shallow threshold with a short evaluation time for testing purposes and intentionally provoked an error. Still, a standard alert rule can have more time tolerance and a threshold based on the needs of the team.

In the end, with everything well set, we saw the bot in action, providing the alert as soon as the trigger conditions were met.

Further reading

- https://dev.to/kirklewis/metrics-with-prometheus-statsd-exporter-and-grafana-5145

- https://github.com/uber/cadence/pull/4793/files#diff-32d8136ee76608ed05392cfd5e8dce9a56ebdad629f7b87961c69a13edef88ec

- https://databand.ai/blog/everyday-data-engineering-monitoring-airflow-with-prometheus-statsd-and-grafana/

- https://www.xenonstack.com/insights/observability-vs-monitoring

- https://www.instana.com/blog/observability-vs-monitoring/

- https://acceldataio.medium.com/a-guide-to-evaluating-data-observability-tools-5589ad9d35ed

Index

Packtpub.com

Subscribe to our online digital library for full access to over 7,000 books and videos, as well as industry leading tools to help you plan your personal development and advance your career. For more information, please visit our website.

Why subscribe?

- Spend less time learning and more time coding with practical eBooks and Videos from over 4,000 industry professionals

- Improve your learning with Skill Plans built especially for you

- Get a free eBook or video every month

- Fully searchable for easy access to vital information

- Copy and paste, print, and bookmark content

Did you know that Packt offers eBook versions of every book published, with PDF and ePub files available? You can upgrade to the eBook version at packtpub.com and as a print book customer, you are entitled to a discount on the eBook copy. Get in touch with us at customercare@packtpub.com for more details.

At www.packtpub.com, you can also read a collection of free technical articles, sign up for a range of free newsletters, and receive exclusive discounts and offers on Packt books and eBooks.

Other Books You May Enjoy

If you enjoyed this book, you may be interested in these other books by Packt:

Data Literacy in Practice

Angelika Klidas, Kevin Hanegan

ISBN: 9781803246758

- Start your data literacy journey with simple and actionable steps

- Apply the four-pillar model for organizations to transform data into insights

- Discover which skills you need to work confidently with data

- Visualize data and create compelling visual data stories

- Measure, improve, and leverage your data to meet organizational goals

- Master the process of drawing insights, ask critical questions and action your insights

- Discover the right steps to take when you analyze insights

Data Wrangling with R

Gustavo R Santos

ISBN: 9781803235400

- Discover how to load datasets and explore data in R
- Work with different types of variables in datasets
- Create basic and advanced visualizations
- Find out how to build your first data model
- Create graphics using ggplot2 in a step-by-step way in Microsoft Power BI
- Get familiarized with building an application in R with Shiny

Packt is searching for authors like you

If you're interested in becoming an author for Packt, please visit `authors.packtpub.com` and apply today. We have worked with thousands of developers and tech professionals, just like you, to help them share their insight with the global tech community. You can make a general application, apply for a specific hot topic that we are recruiting an author for, or submit your own idea.

Share Your Thoughts

Now you've finished *Data Ingestion with Python Cookbook*, we'd love to hear your thoughts! Scan the QR code below to go straight to the Amazon review page for this book and share your feedback or leave a review on the site that you purchased it from.

`https://packt.link/r/183763260X`

Your review is important to us and the tech community and will help us make sure we're delivering excellent quality content.

Download a free PDF copy of this book

Thanks for purchasing this book!

Do you like to read on the go but are unable to carry your print books everywhere?

Is your eBook purchase not compatible with the device of your choice?

Don't worry, now with every Packt book you get a DRM-free PDF version of that book at no cost.

Read anywhere, any place, on any device. Search, copy, and paste code from your favorite technical books directly into your application.

The perks don't stop there, you can get exclusive access to discounts, newsletters, and great free content in your inbox daily

Follow these simple steps to get the benefits:

1. Scan the QR code or visit the link below

https://packt.link/free-ebook/9781837632602

2. Submit your proof of purchase
3. That's it! We'll send your free PDF and other benefits to your email directly

.mcontent.com/pod-product-compliance
Source LLC
urg PA
110090326
B00064B/4586

www.in
Lightnin
Chambe
CBHW0
40690